T0265758

Four Ways of Thinking

DAVID SUMPTER

Four Ways of Thinking

A Journey into
Human Complexity

FLATIRON
BOOKS
NEW YORK

FOUR WAYS OF THINKING. Copyright © 2023 by David Sumpter. All rights reserved. Printed in the United States of America. For information, address Flatiron Books, 120 Broadway, New York, NY 10271.

www.flatironbooks.com

Library of Congress Cataloging-in-Publication Data

Names: Sumpter, David, 1973– author.
Title: Four ways of thinking : a journey into human complexity / David Sumpter.
Description: First U.S. edition. | New York : Flatiron Books, 2024. | "Originally published in Great Britain in 2023 by Allen Lane, part of the Penguin Random House group of companies." | Includes bibliographical references. | Identifiers: LCCN 2024005252 | ISBN 9781250806260 (hardcover) | ISBN 9781250806284 (ebook)
Subjects: LCSH: Thought and thinking.
Classification: LCC BF441 .S839 2024 | DDC 153.4/2—dc23/ eng/20240311
LC record available at https://lccn.loc.gov/2024005252

Our books may be purchased in bulk for promotional, educational, or business use. Please contact your local bookseller or the Macmillan Corporate and Premium Sales Department at 1-800-221-7945, extension 5442, or by email at MacmillanSpecialMarkets@macmillan.com.

Originally published in Great Britain in 2023 by Allen Lane, part of the Penguin Random House group of companies

First U.S. Edition: 2024

10 9 8 7 6 5 4 3 2 1

To Lovisa

Anything sensible in this book is due to you.
I take full responsibility for everything else.

Contents

Four Ways　　5

Starting the Journey　　10

CLASS I: STATISTICAL THINKING

Bright young fellows　　15

Very average friends　　19

A likely answer　　25

The power of stats　　29

Twelve extra years　　32

How do you take your tea?　　35

A happy world　　41

The happy individual　　48

Angry old man　　54

The forest and the tree　　60

More than this　　64

CLASS II: INTERACTIVE THINKING

The cycle of life　　69

Rabbits and foxes　　72

Social chemistry　　78

The social epidemic　　81

More than the sum of its parts　　91

Start a fitness craze　　96

A third law　　100

Cellular automata　　103

Contents

The art of a good argument 111

Top down, bottom up 120

CLASS III: CHAOTIC THINKING

Always knowing the next step 125

Nudge 129

El Farol 133

The chocolate cake of chaos 137

The mistake 143

The butterfly effect 146

The night sky: part 1 150

The night sky: part 2 153

The perfect wedding 158

Cellular chaos 162

A message from B to C 167

Information equals randomness 172

Twenty questions 177

A good listener always asks questions 182

Entropy never decreases 185

Living in a distribution 189

Word games 194

Taking the high road 198

A sea of words 202

CLASS IV: COMPLEX THINKING

The World Congress 211

The matrix 215

Four people in a car 219

Only as complex as its shortest description 221

The streets of London 226

I, II, III, IV   230

All of the life   235

The hard edges of social reality   246

A person is a person through other people   254

Here it is!   258

It's complicated   262

Almost always complicated   263

Who am I?   266

A life in short scenes   272

An indescribable explanation   279

The least words are the deepest   285

Four ways   287

A worthy life   289

Acknowledgements   292
Notes and References   293

Four Ways

The one thing that we just can't stop ourselves from doing is thinking.

Every second of every hour, our thoughts are there. Sometimes instructing us. Sometimes encouraging us. Sometimes telling us we can do better. They analyse what we did in the past and tell us what to do in the future. On and on they go, attempting to understand the world around us.

Yet we seldom consider the ways in which we think. We don't analyse which thought processes get it right and which of them mislead us. Far too often, we fail to think about how we can best shape our thoughts.

We do think about how to take care of our bodies . . . or we at least try to. We decide to go to the gym or go on a diet. We ask ourselves how we can be more motivated or hold the promises we make to ourselves about being healthier. We declare that we need a break from work and talk about how we want to reduce the stress in our lives.

But we seldom take the time to pause, look around and ask ourselves whether or not the way we think about our lives makes sense.

Science and mathematics are, in large part, about finding better ways of reasoning. It is easy to miss this point when we watch a documentary about the origins of the universe, the wonder of the natural world or the structure of our brains and bodies. Science appears, on the surface, to be about facts. But it isn't. Not really. For many scientists, myself included, our primary mission is to shape our own minds in a way that gets us closer to the truth. The facts uncovered by those minds are secondary to us.

This book describes four ways of thinking. Four ways of getting nearer to the truth.

The origins of the four ways approach can be traced back to an article written by twenty-four-year-old former child prodigy and distinguished theoretical physicist Stephen Wolfram in 1984. At that time, Wolfram was working on a set of esoteric mathematical models known as cellular automata. By running computer simulations of these automata on his shiny new Sun workstation, Wolfram had been able to systematically classify the types of patterns they produce. He hypothesized that every process, biological or physical, personal or social, natural or artificial, lies in one of only four classes of behaviour he had observed in his computer simulations. Everything we see or do could be classified as either (I) stable, (II) periodic, (III) chaotic or (IV) complex.

Stable systems are those which reach and stay at an equilibrium. Think of a domino rally, a line of dominoes stood up on end, one after another. When the first one is given a small push, they topple down and come to rest in a line on the floor. Stability. Other examples include a ball rolling down a hill to come to rest in a valley. A pestle and mortar grinding spices into a stable mix of flavours. A dog sleeping peacefully after a long walk.

Periodic systems are those which exhibit repeating patterns. Walking, cycling or riding a horse are periodic motions of, respectively, our feet, the wheel of the bike and the horse's legs. Periodicity is the evenly spaced lines of waves arriving on a beach. It is the fast up-and-down motion as a chef's knife slices the vegetables into equally sized pieces. It is our daily routines: breakfast, work, lunch, work, dinner, TV, sleep and repeat.

Chaos is our inability to predict if it will rain tomorrow or not (in the UK, at least). It is the roll of a dice. The flip of a coin. The spin of the roulette wheel. It is the wild bubbling of boiling pasta water as the molecules vibrate and spin at random throughout the pan. It is chance meetings, when sliding doors open or shut.

Complexity can be found throughout our society. The transport of goods and services across the world. The rise and fall of civilizations. The structure of government and large multinational organizations. But it is also found nearer to home. In our relationships with friends and family, where we can feel both love and frustration at the same time. Complexity also lies inside us. It is the firing of billions of neurons in your brain. It is your personal story, of how you came to be where you are today.

Sometimes our actions move us between classes. Take, for example, disagreements and discussions. I have to admit that I am the sort of person who always wants to get to the bottom of everything and can be quite ruthless in my pursuit of the 'right' answer. If I feel I don't understand an issue or someone adopts a position I don't agree with, then I'll often want to debate it further to find out where the truth lies.

But my love of arguing and debating has at times caused me a few problems, not least with the people I live and work with, who quite understandably don't want to always spend their time going into the ins and outs of absolutely everything.

So, in order to have less pointless arguments, I used Wolfram's theory to establish that there are only two worthwhile arguments: class I arguments, on their way to a stable resolution; and class IV arguments, where important new ideas are discussed but might never be resolved. Class II arguments – recurrent bickering over the same contentious point – and class III chaotic back and forths where we talk over each other are to be avoided.

With this classification in hand, it is easier to identify what class of argument I am involved in. Then, I think about how I might move from class II to class I, or from class III to class IV. I can also think about how I can make my class I arguments converge to stability more rapidly, just like a super-efficient pestle and mortar grinding out the truth.

Notice how thinking in this way changes your perspective: away from the activities themselves and instead offering a view of them from above. Wolfram's classes allow us to think about our overall approach to a wide range of, on the surface, very different challenges.

In 2002, Wolfram published his magnus opus, *A New Kind of Science*, where he proposed an approach to science based around his cellular automata models. The theory was extensive (the book weighs 5.6 pounds and is 1,192 pages long), and the book makes bold claims about how studying cellular automata is a path towards a deeper understanding of biological life, the physical universe and pretty much everything else besides. But Wolfram gave few practical details about how cellular automata provide genuine insight into the real, complex world we live in.

This lack of practical insight has meant that Wolfram's work has never been taken seriously by the scientific community at large. Nor have his ideas made it into the public consciousness. When I searched on Wikipedia for Wolfram's work, I found a page dedicated solely to the mathematical properties of cellular automata. Wolfram's classes have remained abstract, disconnected from reality.

What Wolfram didn't do, and I am going to do now, is to show how his four classes can be used to shape and clarify how we think about the world. These four classes of thinking are not at all abstract. In fact, they are incredibly useful in everyday situations. My approach here is less a New Kind of Science, as Wolfram called it, and more a new way to convince your friends to go jogging with you. A new way of discussing contentious issues with your partner. A new way of controlling chocolate-cake addiction. It is a new way of understanding why we can feel left out at parties. It is even a new way of looking at yourself, as the unique, complex individual that you are.

To find these new, more practical ways of thinking, I have expanded on Wolfram's classes in the four sections of this book.

The way I see it, Class I is statistical thinking. When should you believe the numbers, and when should you be sceptical? Even more importantly, how should you interpret advice from scientific studies about eating and exercise, or about happiness and success? While data and statistics are key to understanding society as a whole, I show that they are seldom as important to *you* as an individual as headlines might lead us to believe.

How then can we find greater fulfilment in life? This question takes us in to Class II interactive thinking: uncovering the secrets of our social world. How can we build a constructive group dynamic? How can we change the way we communicate to resolve disagreements? I explain how to better understand the effect of our own actions on others and how to deal with our feelings when others have treated us badly. Improving your relationships is easier than you might think.

There is a catch, though. The harder we try to control our lives, the more unpredictable they become. Our failed attempts to get ourselves back on track, in a world where it is impossible to know and control everything, often create disorder and randomness. Class III, chaotic thinking, helps you decide when you should try to stay on top of things and when you should let go.

The more complicated a problem is, the more difficult it is to solve. But what does it mean to say that something is complex? I answer this question by arguing that a system is only as complex as its shortest description. By developing an ability to summarize our social situations, our worries and our thoughts in a concise manner, we capture their essential form. Unlike the first three classes, which are very much about solving everyday problems, class IV, complex thinking, is more focused on introspection and self-reflection. It is about finding the stories which help us to better understand ourselves, as well as those around us.

The progress through class I to class IV takes us through the last one hundred years of scientific thinking. It takes us into the minds of some of the scientific heroes (and anti-heroes) who have shaped it. It takes us inwards, into ourselves, and outwards into the world we create together. It takes us from everyday mundane questions about doing housework to the deepest possible questions about what makes us who we are.

And it starts, now, with a young PhD student embarking on a journey of discovery . . .

Starting the Journey

I step off the Greyhound bus into the hot New Mexico sun. The year is 1997 and I am twenty-three years old. It is my first time in the United States. Admission to the Santa Fe Institute's summer school in Complex Systems was highly competitive, but a reference from my PhD supervisor has secured me a place. He had himself previously participated in one of the institute's invite-only research meetings, mixing with the big names. Santa Fe, my supervisor had told me, brings together the brightest minds in physics, economics, biology and mathematics. Their aim: to find a unified approach to complex systems. One which would unite their respective fields and answer fundamental questions. It was to be a new kind of science.

The aim of the summer school was to pass the fast-accumulating knowledge on to the next generation. The participants, other PhD students and young researchers, would be housed for the next four weeks in the dormitories of a small liberal arts college a short distance away from the institute itself. There are lectures every morning and a chance to work on joint projects, under the supervision of the institute's researchers, in the afternoons. In the evenings I would mix with students from all over the world and from a wide range of different academic backgrounds.

'You are going to have an amazing time,' my supervisor had told me. 'Talk to everyone. Try to just soak it all in. It might at first appear that everyone else knows more than you. But they often know less than they let on. So, don't be scared to ask silly questions. You never know what answers you might get.'

It takes me a while to find my way to the institute itself, and when I do finally arrive I realize that I am late for the introduction. As I try to find a seat in the back row of the lecture theatre, the scientific coordinator, Erica Jen, is already in full flow.

'Our aim is to teach you a different way of thinking,' she tells the room full of wide-eyed students, 'but to get there we are going to need to cover a lot of ground. The last hundred years has seen a massive change in how scientists approach their work. We want to make sure you know that history, but we also want to give you an approach that you can't find anywhere else.'

She explains how the lectures will start with the basics, how to handle data and draw reliable statistical conclusions. We will then look at interactions: how predators affect the balance within ecosystems, how neurons signal to each other within our brains, even how our human society changes over time. Then we will learn about the role of chaos and randomness. Why is it so hard to know what is coming in the future? Finally, she says, we will look at the biggest question of them all: what is complexity? What does it mean when we say that we live in a complex society or culture?

Researchers in Santa Fe are looking for mathematical models of our brains and to simulate our social interactions, she tells us. They are trying to find the fundamental dynamics of biological life. The scientists here are great minds who have already proved themselves in their own specialized academic field (many of them have been awarded Nobel prizes) and are now coming together to shape the future of scientific thought.

'These four weeks will be a journey into complexity. They will change the way you think for ever.'

Class I: Statistical Thinking

Bright young fellows

After Dr Jen's introduction, we are shown to the dormitory we will be staying in for the next four weeks.

By the time I find my assigned room, my new roommate, Rupert, has already unpacked. He has taken the bed to the left of the room, nearer to the window, and placed out neat piles of scientific articles and handwritten notes on the only desk.

Rupert tells me he is working on a PhD in economics at the University of Oxford. 'I suppose that's why they put us in the same room,' Rupert ventured, after he found out that I was a fellow Brit.

'Would have been nice to have been sharing with one of those bright young Harvard fellows, really. Expand one's horizons a little . . .' he said, smiling. 'But you'll do . . .'

Rupert had also been sent to Santa Fe by his supervisor, but the instructions he had received were different. He had been told to 'find out what is going on over there' but not to let himself get carried away. This suited him, since he was only vaguely interested in 'all this complex systems malarkey': it clearly wasn't something he planned to waste too much time on. The aim was simply to attend the lectures, try to pick up the essentials, and in the afternoons, he would be studying here, in our shared room. That was why he needed the desk, he explained. And he would, if I didn't mind, prefer not to be disturbed too much.

He didn't seem so impressed with Dr Jen's introduction. 'Typical American salesperson. Hyping it up, as usual,' he said.

Nor did Rupert seem keen to engage with and learn from the other students. Not all of them, in any case. 'I'm sure there will be lots of different types of researchers here,' he said. 'Biologists, historians, sociologists, and the like. There might even be some philosophers. Who knows when you come to a place like this . . .'

15

'They will all be very excited by the idea of "complex systems",' he went on, using his fingers to add quotation marks to his own words. 'I mean, this is a bit like a summer holiday for them all, isn't it? A chance to get away.'

'But,' he warned, 'you and I need to keep our heads.'

Rupert told me he was worried that many of the other participants wouldn't have the same sort of rigorous background as us. There were some basics that the others were probably lacking. And it was our job to enlighten them. For Rupert, it was less about asking questions, as my supervisor had told me, more about gently educating our fellow students.

'We represent something here,' he said. 'We are the rational ones. We are the ones who stand for the use of data. I bet a lot of the other students don't even know basic statistics.'

With that, he sat down at the desk and started sifting through his articles. The conversation was, apparently, over.

I headed out to find some of the other summer-school participants. They surely couldn't all be like Rupert.

I soon found an American theoretical physicist standing in the corridor who introduced himself as Max. I asked him if he knew where the nearest pub was, hoping he might take me there.

He told me that pubs are called bars in the United States or quite often sports bars. He also knew the perfect place and would be very happy to accompany me there.

As we sat down in the bar, Max explained that Americans need continual stimulation and pointed towards the array of TV screens on the walls around us. They can't do just one or two things at a time – like drink beer and talk – they need to see basketball or football too, and, not only that, they need music blasted out during the build-up to every play and the TV screen has to be filled with statistics about the players. I told Max that normally there aren't TV screens in pubs in the UK and, if there are, they are usually switched off.

'It's just a question of time until you have the same,' Max said. He explained that the evolution of American society could be modelled in terms of ever-increasing entropy.

'You know all about entropy, I suppose?' Max asked, not waiting for me to answer. 'It was American researchers after the war who invented the methods for handling, processing and understanding information,' he said, 'and now we are using our skills to feed entropy to the masses.' He smiled knowingly.

I returned a nervous smile, making a mental note that entropy was something I should know more about, and sooner, rather than later.

Max, it turned out, didn't just know about sports bars and entropy. He knew about everything. He was a postdoctoral researcher in statistical physics at Stanford, having completed his PhD at Princeton. I told him about Rupert's lack of enthusiasm for the coming course. Max told me that Rupert was the one in need of an education. Oxford and Cambridge are stuck in the past, he said; they had failed to grasp the importance of chaos and non-linearity (more terms I was unsure about). Oxbridge does valuable foundational work, but it is far too conservative, Max continued. It provides theoretical respectability to the scientific status quo.

That is why Santa Fe is so important, he went on to explain. It isn't that all the best work is being done here, exactly – it was mainly taking place in Princeton and Stanford – but Santa Fe was the meeting point. He listed names: Philip Anderson, Murray Gell-Mann, Kenneth Arrow, Brian Arthur, Chris Langton and Stephen Wolfram. Half of them had won Nobel prizes; the other half were considered maverick geniuses. They had all been here. And the Europeans were starting to take note. 'Rupert will find out soon enough,' Max said.

By now, our table at the sports bar had filled with other summer-school participants.

At the other end of the table, Antônio, an ecologist from Brazil, was dominating the conversation. He was talking in rapid sentences about his new theory of speciation and ecological niches. Eventually, Madeleine, a biologist from Australia, who had clearly had enough of Antônio's mini lecture, suggested that everyone should introduce themselves properly.

We went round the table. Sitting quietly next to Madeleine was Zamya, a philosophy student from France who was attempting to connect the postmodern writings of Jacques Derrida to the work of Ludwig Wittgenstein. Next to her was Alex, from Austria, who had just bought a round of beers and told the others that he was working on chaos theory in chemical reactions. Esther, a Scandinavian computer scientist, said she was starting a study on the network structure of the World Wide Web. I didn't quite follow everything they said about their research – I had no idea who Derrida or Wittgenstein were – but I smiled and told them I was an applied mathematician, looking for a problem to apply my maths to.

After the others had told us about their research interests, Madeleine smiled and said in a broad Australian accent, 'Well, your research all sounds very impressive, but I study the most important thing of all: how ants build trail networks. The most complex system of the lot!'

Antônio responded by starting up again, explaining to Madeleine that ants were a keystone species. The beer mixed with the words inside my head, with the sports commentary, with my efforts to process everything that Max and the others had said, with the feeling of being here in Santa Fe.

Rupert was right: this was a very mixed group of backgrounds and scientific disciplines. Not the usual group of geeky mathematicians I had met on courses at university but people from all around the world. The brightest PhD students in philosophy, biology, chemistry, physics, economics and computer science.

I couldn't imagine any other place I would rather be.

Very average friends

Let's leave my new friends in nineties Santa Fe and return to London today.

It is an average cloudy day in April in the capital, 15°C with a light rainfall. On average, it took London commuters forty-two minutes to travel into their job, for which they receive a median income of around £40,000. When they get home this evening, they will spend an average of 183 minutes watching TV (down from a high of 242 in 2011). Around 51 per cent of Londoners will use social media more than once today, 2 per cent will eat the recommended five vegetables per day and 64 per cent of them will drink alcohol this week. The heterosexual couples among them will typically have sex once a week for a median time of 7.6 minutes. The gay male couples have slightly more frequent sex, one and a half times a week, while modern data on lesbian couples is harder to find. During these Londoners' lives, which will last eighty years on average, they will have 1.6 children. If asked how satisfied they are with their lives (all things considered) on a scale of one to ten, they will answer 6.94, on average.

I could easily write several pages listing statistics and outlining the results of studies about Londoners or, for that matter, about people living anywhere in the world. There's the Office for National Statistics; Our World in Data; Gapminder; the World Bank; national census bureaus; Pew Social Media Reports; Gallup; OECD economic insights; the World Happiness Report; as well as countless other university research surveys documenting our health, welfare, happiness and behaviour. And the statistical relations found in all this data not only inform the decisions made by governments, companies and other organizations but also influence the decisions we make as individuals. We follow recommendations from scientific studies concerning everything, from what we should eat and how often we should exercise to how we can get the most satisfaction from life or best study for exams.

The challenge, when applying statistical thinking to our own lives, is not just knowing what we can say with data but also being clear about what we cannot say. Which of the many scientific studies out there really apply to us as individuals? Do the statistics we read about imply causation, or are they chance correlations? How much should we let statistics and data influence how we see the world? When should we ignore the numbers and use other tools instead?

Before we can answer these questions, we need to start by taking a quick trip into the foundations of statistics and measurement – because it is only after we have an understanding about how statistics are used that we can be more critical about the ways in which they are sometimes misused.

To start our journey, consider how – simply by listing London's average numbers – I was able to paint a picture of a city and its inhabitants which most of us would recognize: the weather, the commute, the payslips, the lifestyle choices, the sex lives. Each of the numbers gives part of an overall impression of London life. The average is the most basic and powerful statistic of them all. Averages tell us the truth about a city.

Statistics also tell us about smaller groups of people. Throughout this book I am going to use the lives of ten friends living in London to illustrate different ways of thinking. These ten people are entirely fictional, but instead of introducing them by describing how they look or writing about their jobs, I am provide some (also fictional) statistics about them in the table on the next page.

If I were to introduce these characters in text, I might write something like 'Nia grabs an oat milk latte on her way to the office in central London, and her assistant always brings her another one at ten o'clock sharp' or 'Jennifer is the eternal student, working part-time to finance her studies. Her idea of luxury is tucking into a jar of pickled gherkins in front of a Netflix series.' The numbers don't have the same colour as words do, but it is striking how they

Name	Age	Yearly Income	Oat milk lattes drunk last week	Likes pickled gherkins?
Antony	34	£12,000	7	Yes (1)
Aisha	31	£36,000	12	No (0)
Charlie	29	£52,000	0	Yes (1)
Becky	29	£23,000	0	No (0)
Jennifer	28	£22,000	0	Yes (1)
Richard	36	£62,000	0	No (0)
Nia	35	£106,000	15	No (0)
John	34	£40,000	0	Yes (1)
Sofie	31	£31,000	5	No (0)
Suki	30	£34,000	0	No (0)

also allow us to gain an impression of individuals. We imagine their jobs, their lifestyles and their particular tastes for pickled foods.

The numbers also tell us a lot about the group as a whole. Their average, or mean, age is

$$\frac{34+31+29+29+28+36+35+34+31+30}{10}=31.7$$

Richard, John, Nia and Antony are a little bit older. Becky, Jennifer and Charlie are a bit younger. But they were (on average) born in the early 1990s, making it reasonable to label this group as millennials.

When comparing incomes, it is common to use the median, rather than the mean. The median is calculated by first writing all the incomes in ascending order, as follows

£12,000, £22,000, £23,000, £31,000, £34,000, £36,000, £40,000, £52,000, £62,000, £106,000 then noting that the middle two incomes are £34,000 and £36,000. Taking the average of these two gives a median of £35,000 per year. This is slightly below the median for London as a whole, but given that most of these friends are at an early stage in their careers, we would still see them as relatively well off. Some of them are probably struggling to get on the property ladder, but none of them are poor. We might wonder how Antony, on £12,000 a year, can afford to drink one latte per day, but one thing I haven't mentioned yet is that Antony is married to Nia, who earns the most. Overall, this is a group of friends who have enough money to get by and a range of opportunities in life awaiting them.*

There is no hard-and-fast rule about when to use the median and when to use the mean (when statisticians say 'the average', they mean the mean, not the median). In the case of the friends' ages, the mean makes most sense because the variation in ages is quite small. For incomes, the median makes more sense because Nia's £106,000 per year skews the value of the average upwards. According to Forbes, London is home to sixty-three billionaires. If we include these super-rich in a mean income calculation, it becomes much larger than the median (often by 25 to 50 per cent for incomes in big cities), making the rest of us feel poorer. The decision whether to use the mean or the median is therefore a question of deciding what we want to highlight in the data. Using the median allows us to ignore those few and far between billionaires.

An extreme example of the difference between mean and median is seen when we look at the oat milk lattes drunk column. The median here is 0 (the majority don't drink any), but the mean is 3.9. Both the mean and the median are needed to sum up our group of friends: it would be equally wrong to say that they don't like lattes as to say they drink nearly four a week!

The distinction between mean and median illustrates that there

* For this section and others containing mathematics, I have created an online lesson going into more detail of mean, median and proportions. See https://www.fourways.readthedocs.io/

are often multiple correct ways of using statistics to describe data. But does this imply that anything goes when it comes to using numbers?

No: there is good statistical practice and bad statistical practice. But how can we be confident that, for example, when calculating the mean age, we are following good statistical practice by adding up the ages of the ten friends and dividing by ten? I used a method all of us have learned in school, but why is it correct? Asking critical questions like this, about the very fundamentals of the way we measure the world, is the key to statistical thinking.

Let's follow this line of critical thought by looking more closely at the data from the gherkin question. The 'Yes' and 'No' answers can be represented as 1 for 'Yes' and 0 for 'No'. Let's write their answers out again, with a 1 if a person likes gherkins, a 0 if they don't.

Antony	Aisha	Charlie	Becky	Jennifer	Richard	Nia	John	Sofie	Suki
1	0	1	0	1	0	0	1	0	0

What is the best estimate, from this data, of the frequency of millennial Londoners who like pickled gherkins?

Intuitively, it feels like the correct answer is $4/10$ or 40 per cent. If we take the average of all the 1's and 0's in the table above, we get exactly this answer:

$$\frac{\left(1+0+1+0+1+0+0+1+0+0\right)}{10} = \frac{4}{10}$$

How do we know this is the correct answer? Imagine, for instance, that some of the friends objected to using the average, employing some admittedly quite dubious arguments. Antony might claim we should give extra weighting to the answers of those who were asked first because 'they are the originals'. He adds up $2 + 0 + 2 + 0 + 2 = 6$ for the first five and $0 + 0 + 1 + 0 + 0 = 1$ for the last five, and estimates the proportion to be $\left(6+1\right)/15 = 7/15$.

On hearing Antony's argument, Aisha counterclaims that it's

better to ask five people and ignore all the others. She just looks at the answer of every second person and finds that, out of this group, only one person (John, as it turns out) likes pickled gherkins and concludes that the proper proportion is 1/5. Finally, Charlie says, 'Hey guys. Let's just listen to the first person and accept what they say as true. It will save us arguments later on.'

Charlie proclaims that 'Antony loves pickled gherkins, so everyone loves pickled gherkins!'

Becky throws her hands up in the air. 'I just don't know about gherkins any more. Charlie says one thing, Antony and Aisha say a lot of things I find very confusing. Let's just agree to disagree. We simply don't know anything about whether a person is going to like a gherkin or not.'

Becky is wrong. Well, she is right that her friends should stop arguing. But she is wrong that we can't say anything about gherkin preferences from the data we have collected. Just because a group of friends have lots of different views does not mean that all their opinions are equally valuable.

The challenge, though, is how we convince Becky, Antony, Aisha and Charlie that there is only one correct way of measuring the proportion of people who like pickled gherkins and it is 40 per cent. We know the friends' arguments are dubious, but how can we demonstrate that this particular proportion is the best estimate?

To do this, we need to travel back in time and visit the man who first realized the need to identify the best way of measuring things.

A likely answer

Imagine a scene in a movie. The camera view is from high above a college quadrangle and the subtitle reads 'Cambridge University, England 1912'. We swoop downwards through a window into the smoke-filled room of a student, Ron, who is sitting alone at his desk. The room is a mess, papers and books strewn across his desk and all over the floor. The student clearly hasn't washed or changed clothes for some time. He is writing frantically, puffing on his pipe and occasionally pausing to search for a particular page in a book.

It is only two weeks until Ron's exams. He will face one of the most difficult academic tests not only in England but in the whole world: the final part of the Mathematical Tripos. When Ron left school he was top of his class and he remains among the cream of the undergraduates at Cambridge. He is soon to have his name recorded in the university register on the list of Wranglers, those students with first-class degrees and the best grade in mathematics at Cambridge.

While Ron gains pleasure from showing off his mathematical skills to others, and has no problem arrogantly admitting his own genius, he is not particularly interested in the upcoming exams. In fact, he isn't even studying for them. His mind is on loftier things. The papers surrounding him are not revision notes but scientific articles; both mathematical, including the works of Carl-Friedrich Gauss and the Revd Thomas Bayes, and biological. Charles Darwin's *On the Origin of Species* lies open on the desk. The notes on the floor outline the principles by which animals (including humans) might be 'improved' using breeding and artificial selection.

Our student doesn't have a proper name for the question he is working on. It's a vaguely formed idea, a notion that it must be possible to pinpoint the only correct way of estimating quantities, in biology and in society, among all the other wrong ways. He suspects that everyone, including his professors, are getting it wrong.

To get a feeling for the way Ron is working, think again about the gherkin controversy in the previous chapter.

This is a special case of the more general question that is consuming all Ron's waking hours in 1912: what is the unique best way of using data to make measurements? A mathematician, especially a Wrangler from Cambridge, has to be able to understand why the calculations they make are the best possible way of doing things.

Ron made the argument as follows. Imagine for now that we don't know the exact proportion of people that will answer 'yes' to the gherkin question but we can be sure it has some value between zero and 100 per cent. He would then ask Antony (who suggested 7/15), Aisha (who proposed 1/5), Charlie (who thinks 100 per cent of people like gherkins) to calculate the likelihood of their suggestions, given the data on gherkin preferences.

Let's start with Aisha's suggestion that the probability that a person likes gherkins is 1/5, or 20 per cent. If she is correct, then the likelihood that we got the answer we got from Charlie is 1/5, since he said he liked gherkins. Similarly, again assuming, as Aisha does, that 80 per cent of people don't like gherkins, then the likelihood of Suki's answer is 4/5. We can now write out a table of the likelihood of each person's answer, as follows,

Antony	Aisha	Charlie	Becky	Jennifer	Richard	Nia	John	Sofie	Suki
1/5	4/5	1/5	4/5	1/5	4/5	4/5	1/5	4/5	4/5

The combined likelihood of all the answers is found by multiplying the likelihoods of all the answers together, i.e.

$$\frac{1}{5}\times\frac{4}{5}\times\frac{1}{5}\times\frac{4}{5}\times\frac{1}{5}\times\frac{4}{5}\times\frac{4}{5}\times\frac{1}{5}\times\frac{4}{5}\times\frac{4}{5} = 0.000419$$

Clearly, the probability of this particular sequence of answers is very small, because it is the probability of us getting a very specific sequence of answers. This does not in itself prove that Aisha is wrong: the probability of any sequence of answers is necessarily going to be quite small. Instead, what is useful about this calculation is that it allows us to compare the likelihood of Aisha's proposal to the other proposals.

To see how, let's start by comparing the likelihood of Aisha's estimate to that of Charlie, who claimed that 100 per cent of people liked gherkins. This gives a likelihood of

$$1 \times 0 \times 1 \times 0 \times 0 \times 0 \times 0 \times 1 \times 0 \times 0 = 0$$

There is literally zero likelihood of getting the answers we did, given his suggestion. He is proven wrong as soon as Aisha gives her answer. So, Aisha wins that one. For Antony's estimate of 7/15 we get

$$\frac{7}{15} \times \frac{8}{15} \times \frac{7}{15} \times \frac{8}{15} \times \frac{7}{15} \times \frac{8}{15} \times \frac{8}{15} \times \frac{7}{15} \times \frac{8}{15} \times \frac{8}{15} = 0.00109$$

Antony is less wrong than Aisha, because 0.00109 is larger than 0.000419. But neither of them is as good as the correct estimate – of 4 / 10 – for which we get a likelihood

$$\frac{4}{10} \times \frac{6}{10} \times \frac{4}{10} \times \frac{6}{10} \times \frac{4}{10} \times \frac{6}{10} \times \frac{6}{10} \times \frac{4}{10} \times \frac{6}{10} \times \frac{6}{10} = 0.00119$$

We have a winner! Our value of 40 per cent has the largest likelihood and is thus the estimate we should use.

Moving back to 1912, we let the camera finally come to rest, angled down over Ron's shoulder, focused on the paper on which he is frantically writing mathematical symbols. He looks up from his pages and lets out a large puff of smoke from his pipe. 'That's it!' he exclaims. '*The maximum likelihood.*'

That afternoon, over a hundred years ago, that Cambridge student saw something no one had seen before him – not even great

27

mathematicians like Gauss, Laplace or Bayes. It was a result that was very different from the mathematics his fellow students were attempting to wrangle in adjacent rooms. Their calculations were impressive, but they were unrelated to real-world observations. It was the link between reality and maths that Ron had wanted to find. And the equation he had now written down achieved exactly that. The maximum likelihood told us the proper way to measure everything – from opinion polling of political parties, through the rates at which plants grow to our taste for gherkins and other pickled products.

It took another twelve years for Ron, whose full name is Ronald A. Fisher, to complete the theory and give a name to the maximum likelihood estimate we still use in statistics today. Fisher is a real person and, while I can't be sure that his theory came to him in the way I describe above, we do know that his work evolved from an article he wrote as a final-year undergraduate student. In the article, Fisher showed that calculating the maximum likelihood provides a single, uniquely correct way of measuring not just the average, as we have done here, but the shape of any curve fitted to data.

Today Fisher's work is considered the cornerstone of statistics.

The power of stats

The first week of lectures in Santa Fe were to be given by applied statistician Professor Elina Rodriguez. Her task, she explained on Monday, was to show us how to best use data in practice. Her lectures were driven by examples. She showed us how to estimate the mean and the standard deviation in people's heights; how to measure the strength of a statistical relationship, such as that between smoking and throat cancer. It wasn't quite what I had expected, given Erica Jen's introduction emphasizing revolutionary new ideas, but Rodriguez made sure we had got the basics.

After her lecture on the second day, I sat down at lunch with Esther, the Swedish computer scientist.

In comparison to the others in the group, Esther seemed slightly detached, as if she knew something the rest of us didn't. She had just finished her masters research project under the supervision of Professor Parker, who was scheduled to teach the second week of the summer school. Parker was working at the Institute for Advanced Study at Princeton and was famous for his original mind and ability to use mathematical models to understand real-world systems.

Esther explained that her masters research project analysed connections between people on the fast-growing internet. Parker suspected that there were deep analogies between how the internet grows and how our brains are structured. Both were complex systems, and he was attempting to unpick the fundamental interactions each exhibited. This felt, to me, much more exciting than the statistics we had heard about from Professor Rodriguez in the morning.

We hadn't had time to get into the details, so, hoping to learn more, I looked for Esther again at lunch the next day. She was sitting at a table with Rupert, a little way away from the other summer-school participants. He was performing a calculation on sheets of A4 paper, explaining it step by step. She was nodding as he spoke and occasionally adding notes to the page with her pencil.

'What are they working on?' I asked Max and Antônio, who were sitting together at another table in the dining hall.

'Rupert is going through the statistics he uses in economics with her, showing how to avoid the most common mistakes. How maximum likelihood works. Dangers of confusing correlation and causation. That sort of stuff . . .' Max said.

'She seems very into it,' I observed.

'Is the Englishman a bit jealous of the other Englishman?' Antônio smiled.

'No,' I replied, slightly embarrassed. 'It's just that I thought Esther would know those results already.'

I told the others that her masters supervisor was Professor Parker. He would be taking us on from pure statistical thinking to other, deeper, interactive approaches, I said. Rupert's stuff was too basic for Esther.

Antônio looked at me. 'It isn't that we can just forget about statistics now we are in Santa Fe . . .'

He explained that in his own work, on rainforest dynamics, accurate measurement is invaluable. He was, just like the rest of us, here to learn about interactions, chaos and complex dynamics, which he believed could help explain how ecosystems worked. But there are basics we all need to know, he said. We can't run before we walk.

'I still feel that Rupert is a bit full of himself,' I said.

That's not the point, explained Antônio. In many situations, there is the right way of measuring things, the right way of doing an experiment, or the right way of looking at a problem. 'Whatever you happen to think about Rupert doesn't change that,' he said.

Antônio was about to say more, but Max shushed him.

Esther had stood up and was walking towards us, Rupert close behind her. She held up the sheets of paper they had been writing on and said, 'I think I've got it now.'

Esther explained that she had studied statistics before, but she had seen it more as a way of testing a hypothesis, like whether a drug has an effect or if a fertilizer helped a crop grow faster, but now she saw that it had even more potential. As we collect more data, which is what is now happening with the World Wide Web, she explained to us, we will be able to find more and more patterns in our behaviour. Automating these statistics will be key to classifying and understanding us.

Rupert stood behind her, smiling, apparently pleased with the new convert to his approach.

'You two seem to have it all solved before the lectures have even started,' I remarked.

Esther smiled. 'I have been working with the "Santa Fe approach" for a while now,' she said. 'It was good to get back to basics.'

When she said 'Santa Fe approach', Esther made air quotation marks with her fingers, just like Rupert had done on the first day with 'complex systems'. He sniggered when she used the hand gesture.

But I could see that, unlike Rupert's closed mind, within Esther there lay an openness. Even before our course had properly started, she had identified that the Oxford economics student knew something she didn't and set about extracting that precise piece of information. Line by line, on the A4 sheets she now held in her hands. She had even started thinking about how to put what she had learned to use.

It was this attitude that my PhD supervisor had been talking about. Setting prestige aside and pursuing what we do not know from whomever happens to know it.

Twelve extra years

When the friends in London started arguing about gherkins and Becky threw up her hands, claiming they should just agree to disagree, she was wrong. The average can sum up a group's opinions about gherkins, or anything else, in a single number. Just like the average age told us that the friends were millennials and the median income told us something about their economic status. These estimates are not true for everyone, but it is the best information we have about a group. Given enough data, collected from enough people, we can make powerful, stable and reliable measurements.

Take health advice, for example. The friends in London are now talking about one of their favourite topics: diets. There are so many to choose from. Suki has been trying out the Atkins low-carb diet. Sofie prefers the Mediterranean diet's focus on food and wine. John has been considering following the Palaeolithic hunter/gatherer lifestyle. Richard has been reading about low-fat foods. There are also popular, competing diets that are everywhere online and happen to be promoted by two mixed martial artists. In the green corner we have James Wilks's vegan diet, which features in the Netflix documentary *Game Changers*. And in the red corner we have podcaster Joe Rogan's diet based on wild game, preferably killed by Joe himself, and fresh vegetables. These two compete for our attention, along with the two days' starvation per week, sugar avoidance, vegetarian and various vegan approaches.

Can Suki, Sofie, John and Richard cut through all the opinions and find the diet that is most likely to benefit their health?

This was the question addressed by a comprehensive survey by David Katz and Suzanne Meller published in the *Annual Review of Public Health* in 2014. Their conclusion: the answer depends on what the question means. If the question is whether there is scientific evidence that eating Mediterranean or Palaeolithic food is a lot better for you than, for example, following the Atkins diet or becoming a vegetarian, then the answer is no. Nor does Wilks's supposedly game-changing vegan diet beat Rogan's hunter-gatherer method. There is no significant difference in the health outcomes for groups of people eating these different diets.

On the other hand, if the question is whether there is a general set of guidelines for how you should eat, then the answer is a resounding yes. The science here is very clear. Provided you avoid eating too much processed food and you do eat plenty of whole, unprocessed vegetables and fruit, it doesn't matter exactly what you eat. All the diets listed above, when followed properly, provide these essentials. The key to healthy eating is, as summarized by Katz and Meller, 'Food, not too much, mostly plants.' It is as simple as that.

There are a few caveats of which we should be aware. Low-sugar diets are good at reducing inflammation. Teenagers pursuing vegan diets for rapid weight loss are not always sufficiently aware of what supplementary nutrients they require. The original Atkins low-carb diet, with its focus on red meat, is not environmentally friendly. Similarly, if we all hunted our own food with automatic weapons, as Joe Rogan might like us to do, many wild animals would be extinct within a few weeks. But none of this changes the larger truth: healthy eating is simply a question of eating your greens and avoiding all the bags, boxes and cans of processed food.

The food industry and much of the media don't want it to be that simple. The average supermarket in the United States offers in excess of 40,000 products, the majority of which are processed foods, many of which carry marketing messages claiming health benefits.

These messages exploit the fact that there is no consensus on individual diets, emphasizing that the food is low fat or low carb, without mentioning that the product is highly processed and thus there is no benefit per se in it being either low fat or low carb. The irony is that food that is good for you – fresh fish, meat, fruit and vegetables – often carries no marketing messages whatsoever.

These types of insights into health are established beyond any reasonable doubt. Scientists have conducted large-scale, long-term statistical studies not just on diets but on all aspects of our lifestyles. Elisabeth Kvaavik, now director of the Department of Alcohol, Tobacco and Drugs at the Norwegian Institute of Public Health, studied the lives (and deaths) of 4,886 people from all over the UK over a twenty-year period stretching from 1985 to 2005. She and her colleagues used maximum likelihood to estimate how the rate at which people died depended upon their lifestyles. What she found should be a lesson to us all. Those people who

adopted four unhealthy behaviours – smoking, drinking more than fourteen (twenty-one for men) units of alcohol a week, leisure-time exercise of less than two hours a week and eating fewer than three portions of fruit and veg per day – had a 15 per cent probability of dying during the twenty years of the study. Those who adopted none of these poor health behaviours had a less than 5 per cent chance of dying. Adopting none of the four unhealthy lifestyle factors reduces the risk of death by two thirds (from 15 per cent to 5 per cent). As Kvaavik and her colleagues wrote, 'Those with 4 compared with those with no poor health behaviours had an all-cause mortality risk equivalent to being 12 years older.'

Lifestyle studies really do apply to us as individuals. Healthy eating, regular exercise, drinking less and not smoking all ensure that *you* live longer. Maybe not exactly twelve years longer – it could only be ten, or it might be fifteen – but they do have a measurable and sizeable effect. Statistical thinking about our health works.

How do you take your tea?

Ronald Fisher has worked through the night to write his master-piece, a paper entitled 'On the absolute criterion for fitting frequency curves'. He publishes it in a small university journal. He waits for the recognition that he is sure he deserves.

That recognition doesn't come immediately. Few of his colleagues read his paper, and those who do find it uninteresting. The mathematics is trivial to Fisher's peers and they don't get its underlying message – that there is one correct way to make statistical measurements. For a twenty-one-year-old who since his teenage years has been competing for and winning prizes for his intelligence, this flat reception leaves him despondent.

And things get worse. Ronald Fisher is poor, his father having lost the money he once had. His bad eyesight means that he isn't admitted, as he would very much like to be, into the army to fight in the First World War in 1914, and he has no alternative but to take a job teaching schoolboys. He hates every minute of it. His fellow teachers view him as aloof and unfeeling, and his students become unruly as he fails to engage with them. Any hope of recognition for his research work is fading, and he blames the stupidity of his fellow man. For him, the solution lies in breeding humans to be more intelligent, to lift the average IQ and create a society full of the types of people who he considers to be enlightened.

Just as an angry young man today might find companionship in online chatrooms discussing taboo subjects, Fisher finds it in editing and writing for publications such as the *Eugenics Review*. At meetings he rails that his nation is 'breeding more from the worse than from the better stocks' and that the only way to save humanity is for men with 'scientific insight, above all their intense appreciation of human excellence' to find worthy women and reproduce. He believes that the coming war may even offer a way forward, stating that 'nationalism may perform a valuable eugenic function'. Fisher is lost in his own anger and frustration.

The light for Fisher comes from what initially appears to be a bizarre decision on the part of this absent-minded academic with a tendency to neglect practical details: in 1917, he decides to become a farmer. It is a way of demonstrating his manliness. Excluded from fighting in the war, he believes that he can prove his worth to his English race through the strength and endurance required to work the land. But Fisher's eventual breakthrough and success doesn't come from his hard work: he leaves the running of the farm to his heavily pregnant teenage wife, Eileen, and her older sister, a London socialite, Geraldine Guinness (who Fisher calls Gudruna because of her Norse goddess appearance), who divorces her husband to follow and finance his adventure. Nor does his success come directly from the rather haphazard experiments he carries out on animals, crops and milk. These serve only to waste even more of Gudruna's funds. At one point, despite them living just above the poverty line, Fisher insists they buy a milk homogenizer for £100 (about half the average yearly salary at the time), which they later have no time to use.

His success instead comes because he catches the attention of the director of the Rothamsted agricultural research station, Sir John Russell, who is looking for an oddball mathematician to 'examine our data and elicit further information that we have missed'. An ex-Wrangler living on a run-down farm with two women and a warehouse full of crackpot experimental equipment fits the job description perfectly. Russell offers Fisher a research position.

The scene is now afternoon tea at Rothamsted in 1919. It is by now a beloved tradition, instigated when a Miss W. Brechley became the first female member of the staff, because, as Sir John says as he introduces Ronald to the others, 'no one knew what to do with a woman worker but it was felt, however, that she must have tea'.

Fisher becomes the most enthusiastic participant in these regular gatherings. He squats on the ground slightly below the trestle table at which the others drink their tea. He is dressed in shabby clothes, leaning forward into the conversation, simultaneously puffing smoke over his companions and attempting to waft it away, as if his pipe is not its cause. He speaks slowly and at length about his views on delicate subjects – not least race – ignoring the blushes of the serious young women sitting at the table.

It is in this setting that Fisher asks Dr Muriel Bristol if he could pour her a cup of tea from the pot. She declines, stating that she prefers the milk to be poured first. To which Fisher responds incredulously, 'Nonsense, surely it makes no difference!' Dr Bristol is not swayed, even in the face of cajoling from those around her. She knows she can taste the difference.

Fisher could not stand the possibility that a claim like this, however trivial, could be made without evidence. Together with another colleague, William Roach (whom Bristol later marries), he sets about organizing an experiment.

Everyone in the tea party knew that a test involving two cups of tea could not be sufficient. Luck alone would allow Dr Bristol to be correct half of the time. Roach suggests that they conduct a series of paired tests, offering her two cups of tea at a time and seeing if she could spot the difference in each case. In this case, the probability that she could get lucky twice in a row is ½ × ½, or 1 in 4. Similarly, the probability that she could get lucky three times in a row is 1 in 8; and four times in a row is 1 in 16. This is certainly a good test. It is quite unlikely that she could have passed the test four times in a row by luck alone.

But Fisher wasn't satisfied. As always, he wanted to find the unique best way of measuring things. Dismissing Roach's idea, he asked the tea lady to make eight cups of tea – four with the milk poured first and four with the tea poured first – and then place them randomly on a tray. He then challenged Dr Bristol to identify the four cups in which the milk had been poured first.

'But what in the world is the difference to my way of doing things?' asks Roach, confused. After all, the methods use the same number of cups.

'Well,' replies Fisher, 'if she isn't able to tell the difference, then the probability she will get all four correct is now 1 in 70, much smaller than 1 in 16. It is a much stricter test. In fact, it is the maximally strictest possible test.'

To understand why Fisher is correct (again), consider all of the ways in which the tea lady could place the cups. When she puts the first cup down, she can choose from any one of eight available cups, when she chooses the second cup she is choosing from seven, and so on. This means there are $8 \times 7 \times 6 \times 5 \times 4 \times 3 \times 2 \times 1 = 40{,}320$ ways of placing the eight cups. Some of these arrangements are identical in the sense that they have the same ordering of milk-first and tea-first cups. We can work out the number of these identical orderings by first multiplying $4 \times 3 \times 2 \times 1 = 24$ to find all the ways of putting down milk-first cups and, in exactly the same way, $4 \times 3 \times 2 \times 1 = 24$ to find all the ways of putting down tea-first cups. Then, finally, we take the ratio

$$\frac{4 \times 3 \times 2 \times 1 \times 4 \times 3 \times 2 \times 1}{8 \times 7 \times 6 \times 5 \times 4 \times 3 \times 2 \times 1} = \frac{24 \times 24}{40320} = \frac{1}{70}$$

to give the probability of each particular unique cup ordering. These are illustrated in figure 1.

Out of all the possible orderings, there is only one correct answer and therefore Dr Bristol only had a 1 in 70 chance of being correct on all four cups. That is, unless she is able to tell the difference . . .

And she was. One by one, to everyone's amazement, she picked out the four milk-first cups from the four tea-first cups. Dr Bristol had nailed it.

Fisher was wrong about Dr Bristol. But in the eyes of his colleagues, he had proved something: they realize that this shabbily dressed mathematician can design better experiments than they can. His methods allow them to increase their chance of success *before* they even start their experiment. Ronald A. Fisher's methods became widely adopted at Rothamsted and, over time, biologists and clinicians further afield adopted his approach. It was when Fisher engaged

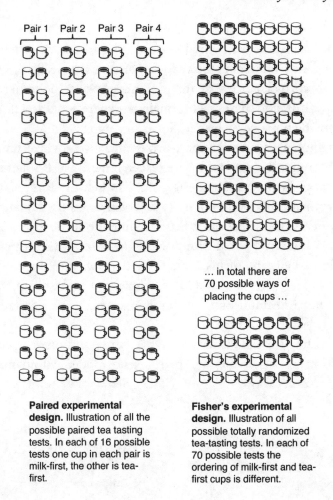

Pair 1   Pair 2   Pair 3   Pair 4

... in total there are 70 possible ways of placing the cups ...

Paired experimental design. Illustration of all the possible paired tea tasting tests. In each of 16 possible tests one cup in each pair is milk-first, the other is tea-first.

Fisher's experimental design. Illustration of all possible totally randomized tea-tasting tests. In each of 70 possible tests the ordering of milk-first and tea-first cups is different.

Figure 1: How to test Dr Bristol's ability to discern whether the milk was put in first.

with very practical problems that his genius was finally recognized, not by the wrangling mathematicians in Cambridge but by biologists throughout the world. As one American statistician later quipped, 'Fisher taught experimentalists how to do experiments.'

It is five years later and our movie's hero, Ronald Fisher, stands in a field of barley explaining carefully to a group of young men and women how they are going to measure yield across different plots. His randomized experimental designs assign each treatment to a random plot, minimizing the chance that any statistically significant result occurs by chance. The uplifting music plays, the camera pans back upwards, and there is a montage of experiments and discoveries, all made possible by Fisher's ideas.

Ronald A. Fisher's scientific work eventually provided him with the recognition he deserved. The research he did at Rothamsted is the basis of how scientists design and conduct experiments today in every field, from microbiology to sociology. His 'fundamental theory of natural selection' was to become the cornerstone of evolutionary biology. His contributions to the theory of statistics were considered by many of his peers to be unrivalled in the twentieth century. He left Rothamsted in 1933 for a professorship at University College London and later returned to his alma mater, the University of Cambridge.

The young Ronald Fisher was right: there is just one correct way of making measurements and treating data, hidden within infinitely many incorrect ways.

A happy world

Fisher's work helped establish the best way to set up experimental or observation studies, in the form of randomized design, and then provided us a framework, in the form of maximum likelihood, for interpreting the results of those studies. Over the next one hundred years, his statistical approach shaped medical trials, psychology questionnaires, sociological surveys and business analytics, and even became the basis for the way social media giants analyse our online interactions. Fisher's work in Cambridge and at Rothamsted is the reason that we now collect so many statistics on all aspects of our lives.

In many situations, we can use the results of research studies to better understand ourselves. We saw this in the way lifestyle choices can be identified as the cause of people living longer.

The success of numbers in finding out about healthy lifestyles does not, though, imply that we need to follow the advice of *every* scientific study we hear about. We need to develop the ability to look at the numbers and question what they really tell us.

Since we last met them, Aisha and Antony have been honing their statistical skills. Becky has been learning how to be a more constructive sceptic. And Charlie has been scouring newspaper and online articles in order to find scientific studies of happiness.

Charlie found an online report called the World Happiness Report. Every year since 2005, the report's authors have analysed the results of the Gallup World Poll, which is carried out in 160 countries (covering 99 per cent of the world's population). The pollsters contact a random sample of people in each country and ask them over a hundred questions – about their income, their health and their family. These include the following question about happiness:

> *All things considered, how satisfied are you with your life as a whole these days? Use a 0 to 10 scale, where 0 is dissatisfied and 10 is satisfied to give your answer.*

The numerical answer can be used as a measure of personal happiness. Answer this question yourself, now, in your head, before continuing. How satisfied with life are you on a scale from zero to ten?

People living in different countries give different answers. I've already given the average for the UK for 2022 in the introduction. It is 6.94, making the UK seventeenth in the world for happiness in that year. The top-ranked country is Finland, with a score of 7.82. In general, Scandinavian and northern European countries are ranked highest. The USA is sixteenth (0.03 points ahead of the UK). China, with a score of 5.59 and at seventy-second place, is roughly in the middle of the table of the countries surveyed. Other mid-ranked countries include Montenegro, Ecuador, Vietnam and Russia. Further down the table, we find many African – Uganda and Ethiopia placed at a hundred and seventeenth and a hundred and thirty-first, respectively – and Middle Eastern countries – Iran is a hundred and tenth and Yemen at a hundred and thirty-seventh. The unhappiest country in the world in 2022 is Afghanistan, with an average happiness score of only 2.40.

In order to better understand differences between countries, Antony plots the average life expectancy of each of these countries against their happiness scores. These are shown in figure 2a. Each circle in the plot is a country. The x-axis shows the life expectancy in the country and the y-axis the average ranking of life satisfaction on the zero to ten scale. In general, the higher the life expectancy of a country, the higher the happiness there.

One way to quantify this relationship is to draw a straight line through the points, showing how happiness increases with life expectancy. For example, imagine that for every twelve extra years which people live in a country they are one point happier. The equation for happiness in this case would then look like this

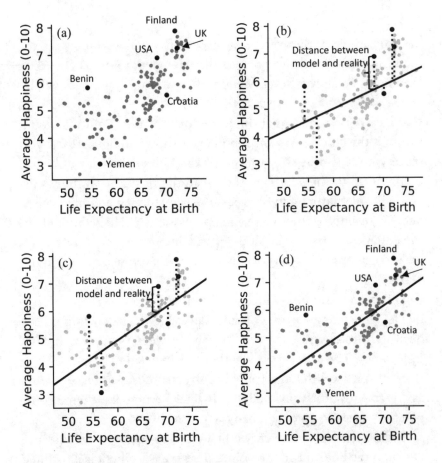

Figure 2: The relationship between life expectancy at birth and average happiness within 136 countries around the world. For details see the World Happiness Report 2019. (a) Each country is represented as a grey circle. Certain countries are highlighted in black. (b) One potential straight line relationship with slope 1/12 and intercept 0 (not shown on the figure), described in the main text, between happiness and life expectancy. The dotted lines connected to the solid line illustrate the distance between the model and reality for specific countries. (c) The maximal likely straight line relationship has slope 0.112 and intercept –2.41, between happiness and life expectancy. (d) Once the model is fitted to the data, we can compare how well it compares to reality in different countries.

$$\text{Happiness} = \frac{1}{12} \times \text{Life expectancy}$$

For example, if the average life expectancy in the country is sixty, then the equation above predicts the happiness to be $60/12=5$. If the life expectancy is seventy-eight, then average happiness will be $78/12=6.5$.

We can draw this equation in the form of a straight line going through the cloud of country points, as shown in figure 2b. If you place your finger at sixty on the x-axis (Life expectancy at birth), follow it up to the solid line then read the position on the y-axis, you will see that happiness is 5. Similarly, starting at a life expectancy of seventy-eight will give happiness of 6.5. The value $1/12$ in the equation above is the slope of the line: for every twelve years we move along the x-axis, we increase one happiness point on the y-axis.

This particular line – which predicts that happiness is one twelfth of life expectancy – is one of many possible lines that might have been used to describe the relationship between happiness and life expectancy. The question is whether this line is the 'best' line? The slope of one twelfth looks roughly correct, but is it, as Fisher would ask us to show, the maximally likely line? Remember, there is only one right answer, amongst many wrong answers.

To find the right answer, we first need to measure the distance between the line and all the points. This measurement is illustrated in figure 2b for the line we saw above, where happiness is predicted to be life expectancy divided by twelve. The dotted lines connecting Benin, Yemen, Croatia, the USA, the UK and Finland to the solid line show how far each of these countries lie from the predictions, i.e. the distance between prediction (the solid line) and reality (the circles for each country).

The line that is closest, on average, to all the points is given by the equation

$$\text{Happiness} = 0.123 \times \text{Life expectancy} - 2.425$$

and shown in figure 2c. The slope of this line is slightly steeper than that in figure 2b and the intercept is negative −2.425 (it was zero for our first equation).

I have stated that the second line (the one in figure 2c) is closer to the points than the first line (the one in figure 2b), but how do I know for sure in a way that would satisfy Fisher? To answer this question we sum up the square of the distances between the line and all the points. For example, the USA has a happiness score of 6.88 and an average life expectancy of 68.3. The first equation (figure 2b) predicts

$$\text{US happiness} = \frac{1}{12} \times 68.3 = 5.69$$

which means that the squared distance between the prediction and reality is $(6.88 - 5.69)^2 = 1.416$. The second equation (figure 2c) predicts

$$\text{US happiness} = 0.123 \times 68.3 - 2.425 = 5.98$$

which means that the squared distance between prediction and reality is $(6.88 - 5.98)^2 = 0.8100$, which is smaller than the value 1.416 for the first equation. This means that, for the USA at least, the second equation is closer than the first.

We repeat the same calculation for every country and for each line and then sum over all the countries. In statistical language, this method is called calculating the sum of squared distances. We consider the best fit line to be the one which has the smallest sum of squared distances. The sum of squared distances for the line in figure 2b is 82.84, while the line in figure 2c has a smaller sum of squared distances of 71.76. So the second line is closer than the first. On the webpage for this book, I go through the steps of calculating these sums of square distances (see the Notes for more details). I also show that the second equation is not only better than the first, it is also the line closest to the data in terms of sum of squared distances. There is no line which is closer on average to the country data points than this one. This doesn't mean it is better for all countries. For example, the line in figure 2b is closer for Croatia. But averaged over all countries, the line is closer to the data.

Antony and Aisha are sitting at their computer admiring the line they have fit through the happiness data when Becky comes in.

'Here we go again,' whispers Antony to Aisha, remembering how Becky told them all to agree to disagree when they discussed gherkins. 'Let's tell her what we have found out before she has a chance to question it.'

Before Becky can speak, Aisha starts explaining that people who live longer and are healthier are also happier. The secret to happiness is a long life, she says. The straight line through the cloud of points is the proof of the relationship.

But this time Becky is also prepared. She has been reading the World Happiness Report from which the data is taken. She tells them that she has found that life expectancy is not the only measurement that is correlated with happiness. When she looked at cross-country data, she also found a straight-line relationship between happiness and Gross Domestic Product (GDP, a measure of economic wealth), and between happiness and the proportion of people who, when surveyed, answered yes to the question 'Are you satisfied or dissatisfied with your freedom to choose what you do with your life?'; how much the average person in the country donates to charity; and perceptions of corruption in a country. Becky tells the others that the single best predictor of happiness, even better than life expectancy, is whether or not people feel they have the support of those around them. Countries in which people are more likely to answer yes to the question 'If you were in trouble, do you have relatives or friends you can count on to help you whenever you need them, or not?' also have higher life satisfaction – they are happier people.

The relationship between happiness and life expectancy, Becky explains, is extremely complicated. There are so many interrelated factors and the straight-line model is highly simplified.

'From this data, we simply don't know whether a person is going to be happy or not. We certainly can't say anything about our happiness as individuals,' she says.

This time, Becky is right.

John Helliwell and his colleagues, who collated the data for the World Happiness Report, emphasize the importance of what they call social foundations in creating a happier world. Happiness arises when we have choices, when the people around us are generous and sociable, when we don't live in poverty and when we are likely to live long lives. But we have to be careful when interpreting these results. We cannot know, from cross-country comparison data alone, which factors *cause* happiness or merely happen to *correlate* with it. We don't know if better healthcare or better social support causes an increase in happiness, or if nations in which people have developed a more positive outlook on life build better healthcare and social support. What we do know is that people in more stable, more prosperous countries with greater social support tend to describe themselves as happier.

Unlike large-scale studies of health, which we looked at earlier, we cannot rely on national questionnaire results to plan our own path to personal happiness. The majority of the factors which Helliwell and his colleagues have studied are not under your control. Just because people in Finland say they are happy, it doesn't imply that moving to Finland will necessarily make you happy. Nor does simply knowing you will live longer guarantee life satisfaction. There are complex relationships between economic development, health services, social security, democracy and freedom of expression which underlie the correlations we see in the data. Simply put, we cannot separate cause and effect when we compare countries.

The happy individual

How can we find the cause of happiness for ourselves as individuals? The friends decide to switch focus from country comparisons to studies which focus on individuals.

One media headline that caught Charlie's attention was 'Yes, you can buy happiness . . . if you spend it to save time'. The accompanying article in *USA Today* explained that psychology researchers had found that people who spend on housekeeping, delivery services and taxis are happier than those who don't. The research in question was conducted by Elisabeth Dunn, Professor at the University of British Columbia, and her colleagues, including Ashley Whillans, Assistant Professor of Business Administration at Harvard. Charlie downloaded the scientific research article, published in the journal *Proceedings of the National Academy of Sciences*, in order to find out what the four friends can learn.

Dunn and her colleagues had first undertaken surveys in the USA, Canada, Denmark and the Netherlands, looking at how people spent their money and how happy they are. The approach was similar to the Gallup poll we saw earlier, but the analysis is of individual people, not countries. This is much more relevant to Charlie and his friends, who are, obviously, people and not countries. The researchers found that people who spent more on time-saving purchases each month have a higher score on the life-satisfaction scale.

When Charlie tells Aisha and Antony about the results, the three friends realize they can use their newly acquired statistical skills to look at the data from the study in more detail. One of Dunn's long-term collaborators, Lara Aknin, Psychology Professor at Simon Fraser University, has created a repository of data from their various studies of the science of happiness. Repositories like these, in which all data is anonymized so that study participants cannot be identified, are a gold standard for psychology research. They allow other researchers, even amateurs like Aisha and Antony, to better understand and verify a study's conclusions.

Aisha and Antony start by looking at data from one of the questionnaires carried out in the USA. Respondents who did not spend money on time-saving purchases had a happiness score of 6.70, on average, while those who did had an average score of 7.22. This difference can't be attributed to chance, since over one thousand people were surveyed.

When they tell their results to Becky, she remains sceptical. 'People aren't all the same,' she says. 'This result doesn't mean that *all* people who spend money to save time are half a point happier than all of those who don't? Does it?'

To answer Becky's question, Antony plots two happiness histograms, one for those who did spend money to save time, and one for those who didn't (see figure 3). These histograms show that Becky's concerns are real: people vary a lot in how happy they are, and the difference between the two groups is small. Those not spending money on time-saving report scores of 5 and 6 slightly more often, and those who do spend money on it score 9 or 10 slightly more often. But both groups have a lot of 7s and 8s.

Aisha explains to Becky the result of a statistical test she has done. She has picked one random person from those who *did* spend money on time-saving and one person at random from those who *didn't* spend money on time-saving and compared their happiness. By repeating this procedure 100,000 times (using a computer), she calculates the proportion of times the person who did spend on time-saving is happier. This proportion is only 55 per cent.

You can think of this result in terms of the answer you gave to the '*All things considered, how satisfied are you . . .*' question I posed earlier. If you currently do not spend any of your monthly budget on time-saving, then this value of 55 per cent gives an estimate of the probability that you would be happier if you did. Likewise, if you currently do spend some of your budget on time-saving, then the probability you would be happier if you didn't make the spend

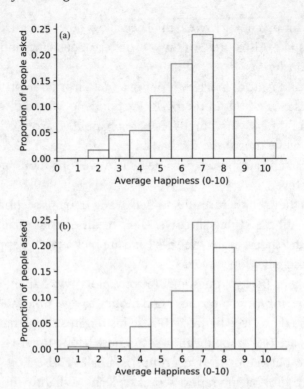

Figure 3: Happiness histograms. Shows the proportion of people who gave each possible happiness score from 1 up to ten for (a) people who didn't spend money on time-saving purchases and (b) people who did spend money on time-saving purchases.

is 45 per cent. It makes sense to spend money on time-saving, but results are by no means guaranteed. If spending money had no effect whatsoever, the probability would be 50 per cent.

In the survey, participants reported their monthly spending on material items, experiences and time-saving services. Antony now calculates the straight line with the smallest sum of square distances through this data. He finds that every $100 spent on time-saving leads to an average gain of 0.31 points on the happiness scale. So, increasing spending from $0 per month to around $300 per month on saving time provides almost one extra point of happiness. The straight-line model also shows that increasing spending on experiences is half as effective as spending on time-saving – with spending on material purchases being only one fifth as effective. This analysis suggests that it is not spending money per se that makes us happier, but that buying time, and to some degree experiences, is much better than buying things. Antony's analysis further shows that spending over $300 has diminishing returns, so it isn't worth investing much more money than this on time-saving.

Charlie is impressed with this analysis. This time he feels he can interpret the results in terms of his own lifestyle choices: spending money on time-saving might turn out not to work for him, as an individual, but it is certainly worth a try. But Becky has a further concern. 'This doesn't prove that spending on saving time *causes* people to be happy,' she says. 'It could also be the case that happier people are more willing to spend money on saving time.'

Questions like this can be answered with good experimental design, as Fisher did in his tea-tasting experiment and as he proposed for testing the growth of different crops. By finding people with more or less similar backgrounds and randomly assigning them to two groups, we can test the effect of a particular intervention.

Dunn and her colleagues conducted such an experiment. The researchers recruited people at a science fair, gave them $40 one week to make a time-saving purchase and $40 another week (either before or after) to make a material purchase. The order in which they could make their purchases was randomized. Aisha has looked at what the sixty participants bought on each occasion and the difference in happiness that resulted. For example, one woman who bought 'eye liner and make-up' one week was much happier the next week when she instead 'took a taxi and tipped the driver'. A man who bought an 'outdoor playset' one week and 'dinner for the family' the next was much happier after the dinner. In some cases, the opposite was true: a woman who bought 'hiking gear' was happier than when she 'got her nails done'. Overall, twenty-six out of the sixty people participating in the study were happier with a time-saving purchase, fourteen were happier with a material purchase and for twenty it made little difference.

To interpret this result, we need to understand two key ideas from statistics: significance and effect size. Statistical significance is a measure of the probability that results of a study would have arisen by chance. Specifically, consider the forty people who were happier after either a time-saving purchase or after a material purchase. If there was no difference between the types of purchases, we would expect twenty of them to be happier after a time-saving purchase and twenty to be happier after the material purchase. If we found that twenty-one or twenty-two were happier after a time-saving purchase, we wouldn't rush to the conclusion that time-saving purchases are definitely better. Small differences like this can be explained by chance.

The question then is whether twenty-six who were happier after making a time-saving purchase out of forty (for whom a purchase made a difference) can be explained by chance. We can test this quite simply: we just have to calculate the likelihood of getting twenty-six or more heads after tossing a coin forty times. This probability is around 2 per cent, which is rather unlikely, and so we can say that the result is statistically significant.

Statistical significance is not the same as effect size. In the study, the median difference in happiness between time-saving and material purchases was 0.167 on a scale ranging from zero to five. This is a small effect on happiness. Aisha found a similarly small effect when she considered what happened when she picked two people at random – one who did spend on time-saving and one who didn't. In 45 per cent of pairwise comparisons, it was the person who didn't spend money on time-saving who was happier. Again, the result is statistically significant (it can't simply be explained by chance). But even if spending money on time-saving had no effect on happiness, in pairwise comparisons, we would expect the person who had spent money on time-saving to be happier 50 per cent of the time (just like we expect to get heads half the time when we toss a coin). When compared to 50 per cent, a 45 per cent effect is quite small.

When we read headlines about research studies, we need to consider statistical significance and effect size, along with causality. The study we looked at earlier about lifestyle and length of life is statistically significant, causal and has a large effect size (twelve extra years). The between-country study of the relationship between life expectancy and happiness is statistically significant and has a large effect size but does not establish a causal relationship. The studies looking at time-saving and happiness are causal *and* pass the statistical significance test, but the effect size is small.

So, next time you read a headline, watch an inspirational TED talk or click on a link promising you happiness, think about these three aspects – causality, statistical significance and effect size. All three need to hold for a study to apply to you.

Angry old man

'I was a bit sceptical when I saw Rodriguez's name and university affiliation . . . some college in upstate New York,' Rupert tells me as we walk to the lecture theatre for the penultimate lecture of the week, 'but she has presented the subject well. And I see that today's lecture is on Ronald Fisher. I have to admit, I'm impressed.'

Rupert explains that Fisher is one of his scientific heroes. Not only had he worked on the design of experiments, maximum likelihood estimation and other aspects of statistical theory, he had also made massive contributions to genetics and mathematical biology. For Rupert, Fisher embodies the assertive attitude required to succeed in modern science.

Sometimes Fisher could be a bit sharp, Rupert tells me. For example, when his ground-breaking book, *The Design of Experiments*, was published in 1934 to a glowing review in the *British Medical Journal*, Fisher took umbrage at an off-the-cuff remark made about how his experiments were conducted at an agricultural field station. He wrote to the journal to give the exact number of pages on which such experiments were described, emphasizing that his results also came from other areas of research, not just farming.

Fisher had a legendary correspondence with a group of fellow statisticians whom he labelled 'the mathematicians', those professors around the world who, Fisher felt, didn't recognize his brilliance. He called them the 'mathematicians' because, in Fisher's view, their work was abstract, while his own was connected to reality. These mathematicians seldom won arguments with him. Fisher was, Rupert explains, a better statistician, a more knowledgeable biologist and a more creative thinker than almost all his peers. One colleague described him as burning 'even more than the rest of us . . . to be original, right, important, famous, and respected'.

As we sit down in the lecture theatre, Rupert summarizes, 'If we all pursue our ideas with the same passion and exactitude as Fisher, then great discoveries can be made. We don't need to get carried away with this mystical complexity stuff they are going to push on us in the coming weeks.'

Professor Rodriguez stands in front of the classroom, waiting for us to quieten down. Only once it is completely silent, and still without saying a single word, does she put up a picture of Ronald Fisher on the overhead projector.

'For many people, Ronald Fisher is a hero,' she begins, 'and he is certainly a good starting point for understanding the statistical approach upon which science is built. But there is another side of Fisher. And it is that side we are going to look at today . . .'

Rodriguez tells us that, at work, Fisher was fiercely competitive and opinionated, shouting down those who disagreed with him. One of Fisher's friends described him as 'eccentric, cantankerous, opinionated, and often vehemently subjective'. At home he was even worse. His daughter and biographer, Joan Fisher Box, witnessed first hand how, sometimes a 'towering rage . . . would bury his wife under its volcanic debris'. When one of his research programmes was shut down during the Second World War, he became angry, cruel, paranoid and began a course of 'sadistic persecution' against his wife. He would 'slap down' his children if they dared to defend their mother.

Some of the summer-school students let out gasps during Rodriguez's description of Fisher. Others shake their heads disapprovingly. But, as soon as she finishes, Rupert puts up his hand and, without waiting, asks, 'What has Fisher's home life to do with his science?'

'Well,' replies Rodriguez, 'the programme that was shut down before the war was focused on eugenics: an attempt to optimally breed both animals and humans. Fisher viewed some human beings as inherently inferior and others as inherently superior. And he wanted to ensure that the individuals he saw as superior had more children.'

As a young man, Fisher believed that the variation he saw in his fellow human beings' intelligence and success was a consequence of racial differences. He thought social classes and nations contained different genetic characteristics and were competing with each other for dominance. During the 1920s and 1930s, in the shadow of the Second World War, Fisher campaigned tirelessly for a law which would allow 'feeble-minded' women to access 'voluntary sterilization'.

'What was most bizarre about Fisher's sterilization campaign,' Rodriguez tells us, 'was that he soon found himself arguing against his own theoretical results.'

Rodriguez encourages us to temporarily set aside the problem with the definition of 'feeble-mindedness' – which we know today isn't a valid medical diagnosis – and even set aside the completely unethical idea of a sterilization programme. She asks us to assume, as unpleasant as it now seems, that Fisher and the other eugenicists of the day were acting in good faith when they thought it possible to label certain people in that way.

The problem with their argument was that they knew, even then, that so-called 'feeble-minded' children did not usually have 'feeble-minded' parents. Nor did all 'feeble-minded' parents have 'feeble-minded' children. This implied that the supposed allele (gene) for 'feeble-mindedness' was recessive; that both the mother and father must carry it in order for the child to be 'feeble-minded'. Fisher's Cambridge colleagues had already proved, in 1915, that eliminating rare recessive alleles took thousands or tens of thousands of generations. Elimination of 'feeble-mindedness' was impossible, even if an effective sterilization campaign had been implemented. We now know there is in fact no gene for 'feeble-mindedness' – intelligence is instead a complex combination of many different genes and environmental factors – but even based on the science of the day, there was no justification for Fisher's position.

'You see,' Rodriguez said, looking directly at Rupert, 'this is why Fisher's character and personality, and even his home life, are important. His proposals on eugenics were not only highly unethical, but he was belligerent in his defence of an idea that was scientific-ally unfeasible. And this was not the only mistake of this kind he made . . .'

After the Second World War, Fisher's interest in human eugenics waned, to be replaced, in the 1950s, by something else that lay close to his heart: smoking. He accepted that there was a correl-ation between smoking and cancer, but he argued that the evidence was not sufficiently strong to establish causation: smokers might develop cancer more often, but that does not in itself imply that smoking causes cancer. His alternative hypothesis was that there was a genetic link between the tendency to smoke and a tendency to develop cancer. The same genes could play a role in both smok-ing and cancer.

At that time, Fisher's hypothesis was difficult to completely rule out. The structure of DNA had only just been established and none of the genetic techniques we have today were available. Fisher set about conducting small-scale studies that supposedly supported parts of his hypothesis. He found that identical twins were more likely to have the same smoking habits than non-identical twins. He dug out previously discarded data showing that smokers who said that they inhaled when smoking were less likely to get cancer than those who claimed not to inhale. None of this work provided any conclusive evidence for his theory, but it continued to cast small elem-ents of doubt on the causal link between cigarettes and cancer.

Despite the best efforts of the tobacco industry, which financed Fisher and other statisticians at the time, the link between smoking and cancer was eventually shown to be undeniable. On balance of vast quantities of evidence, the US Surgeon General now attrib-utes almost half a million deaths per year in the USA to smoking. One possible reason that Fisher's data showed that those inhaling smoke were less likely to develop cancer is that those who have developed cancer feel better about their lifestyle choices by later claiming that they didn't inhale. Fisher had confused correlation with causation. One by one, his arguments fell away, but it was only after Fisher's death, of complications arising from colon cancer in

1962, that the experiments required to conclusively prove him wrong were completed and health warnings were finally included on cigarette packets. It is very likely that the 'genius' statistician's tireless attempts to undermine the facts led to the loss of many lives.

'I have told you these two stories at the start of our journey both as a warning and a way of setting the stage for what we will now learn. We have to do better than Fisher. We have to be better scientists *and* better people,' Rodriguez warns us.

Rodriguez explains that a determined contrarian, like Ronald Fisher, can use their statistical skills to argue against the truth. Such a contrarian can find plausible alternative hypothesis after alternative hypothesis. They can claim that they are being neutral, just raising different possibilities. Fisher's papers on smoking are, like those on 'feeble-mindedness', a masterclass in dismissive bluster. He used his title and academic authority to lend credibility to his claims. He belittled his opponents, claiming that they couldn't follow his arguments because they lacked sufficient skills in statistics. But ultimately, he was defending the wrong position.

In any good scientific research group or community, Rodriguez says, a balance is needed between the contrarians like Fisher and the less individualistic majority who push for consensus. Getting that balance right is something all scientists think about – a lot. We want our hypotheses to be challenged, but we don't want to be paralysed by uncertainty. We want to get as close as possible to the truth, given the limited time and resources we have to collect data and conduct experiments.

Rodriguez reminds us again of Fisher's work in Cambridge before the First World War. This was a young man who, against all the odds, discovered the best possible way to measure things. In the process, he was able to show how statistics contribute to scientific progress. It was a great leap forward in thinking that is easily taken for granted – and Fisher was perhaps more responsible for it than anyone else in human history. It was a remarkable achievement and for this he is one of the great heroes of twentieth-century science.

But what Fisher failed to fully grasp, even as he got older, was that the decisions he made – about what data to collect, about the things that were measured – could introduce a new form of bias. If we focus on analysing data on whether smokers are inhaling, while ignoring the most important data that shows they are dying, or if we introduce a medically unsupported diagnosis of 'feeble-mindedness' to describe people who have never been given access to education, then it doesn't matter how accurately we take measurements.

When it suited him, Fisher neglected causality and failed to consider the small effect size of the studies which supported his own worldview. Instead, he used his skills in statistics to push us towards certain ways of seeing a complex world through an overly simplified lens. He presented his own prejudices – that smoking wasn't bad for us and stupid people should not breed – as objective science.

'In this way,' Rodriguez concludes, 'there is just as much to be learned from Fisher's failures as there is from his successes.'

The forest and the tree

The power of statistical thinking comes from how it measures relationships in data. We can say with certainty that adopting all aspects of healthy living provides us with twelve extra years of life on average. We really should cut down on the booze, eat our veg and get our bodies moving. Statistics also allow us to see how happiness, security, wealth and life expectancy correlate across countries. These are just two examples of countless ways in which medical and sociological studies have been used to inform health and public policy.

But we have also found that statistical thinking is not enough. It is not sufficiently powerful to distinguish correlation from causation. Happier countries are also wealthier countries, but it does not necessarily follow that money buys happiness. Carefully conducted experiments and well-thought-out observations are needed to disentangle these two things, causation and correlation.

Statistics can be abused. An unscrupulous individual can hide the truth by measuring the wrong thing and using numbers to cover their trail. Out of the three evils Mark Twain identified – 'lies, damned lies and statistics' – the last is often the worst. Fisher proved this point, just as thoroughly as he proved that there are right and wrong ways to understand data and conduct an experiment. He used statistics to lie about smoking and cancer. He used numbers to justify a repugnant theory of eugenics.

Fisher's mistakes don't mean that statistics always lie – in practice, they usually unveil the truth and are used honestly by researchers – but they tell us that, used in the wrong way, they can obscure and mislead.

The limits of statistics can take very subtle forms. Even carefully designed experiments often explain only a small proportion of the variance between individuals. So, while it is true in general that people who spend more money on saving time or on helping others are happier (which is another conclusion of the work done by Elisabeth Dunn and her colleagues), this does not mean that it applies to you, as a unique individual. For the happiness study we looked at, a majority of people will *not* benefit from following the advice to spend money on saving time. This doesn't mean it isn't worth considering, but it does mean you shouldn't get your hopes up or get too disappointed if this or other ideas you have read about in a newspaper headline fails to work out for you.

This mistake, of seeing research results as applying to ourselves as individuals when they are really a measure of the group as a whole, is often referred to by statisticians as an example of an ecological fallacy. I like to describe it as confusing the tree (you as an individual) for the forest (the group as a whole).

Before we continue our journey into other forms of thinking, let's pause to look at this limitation more closely. Books, newspapers, social media, YouTube videos and TED talks provide us with access to a vast array of scientific studies of our psychology, our motivations and our personalities. Each of them has its own suggestion about how we can be happier, more successful and more satisfied with our lives. How can you know which of these studies applies to *you* as an individual?

Take, for example, the idea of having a gritty personality. Angela Lee Duckworth's TED talk 'Grit: the power of passion and perseverance', is one of the top twenty-five most watched TED talks of all time. The talk is based on a study that Duckworth and her colleagues conducted on college undergraduates, West Point army cadets and young US spelling-bee competitors. She asked each of the study participants to rank twelve statements – such as 'I become interested in new pursuits every few months' and 'Setbacks don't discourage me' – on a scale of one to five as to how well it described them. These scores, when combined, measure the participants' grit. Duckworth found that students with more grit got higher grades, cadets with more grit were more likely to make it through their first summer training programme and spelling-bee contestants with more grit were more likely to progress to the final of the competition.

Duckworth's study is rigorous, using the statistical methods we looked at earlier when studying happiness to show the relationship between grit and achievement. But . . . what might not be obvious to all the TED talk's 23 million viewers asking themselves if they have got enough grit to succeed is how much of achievement is explained by the desire to see things through and how much is explained by other factors. The answer to the first is very little. Although grit explains around 4 per cent of the variation between individuals, this leaves a further 96 per cent unexplained. There are some gritty trees in the forest who are successful, but that does not mean that you individually can succeed by becoming grittier.

In the last five or ten years, psychologists have undertaken comprehensive meta-studies looking at how aspects of our personality and our psychology and our experiences effect various outcomes in our life. In meta-studies the results of a large number of independent experiments are combined in order to find an overall effect size, and the effects are often found to be small. When grit has been tested in meta-studies, it has been shown to explain only a tiny amount of the variance between students. Similar results have been shown for an idea known as growth mindset, which says that we should emphasize for students that abilities aren't set in stone and that, with work, we can change ourselves. Although widely promoted, well meaning and certainly true in some broad sense, an important question is whether emphasizing growth mindset as part of an educational philosophy in school improves students' exam results. Experimental observations have shown that the growth mindset approach only applies to specific types of students (those who are lower achieving in exam situations) and, even then, it explains only a very small amount of the variation between these students.

Many of the 'inspirational' ideas which permeate our collective consciousness have only very limited application to you as an individual. Positive psychology interventions – like asking participants to write down three good things that happened to them during a day – can be helpful for some, but they explain only around 1 per cent of the variance between people. Another oft-touted, alternative way of measuring study and work skills is emotional intelligence. Again, when general intelligence and conscientiousness (a personality measurement similar to grit) are accounted for, emotional intelligence explains only 3 or 4 per cent of the variance between individuals in their academic performance.

When you watch an inspirational presentation based on scientific smart thinking, like a TED talk, or read about the latest study on how we can become happier, better people, you should always bear in mind the ecological fallacy: as interesting as this result happens to be, it probably does not apply to you. You are a tree. The study is of a forest.

Numbers are essential to understanding humanity, but they are not enough. If we want to know about ourselves and those around us as individuals, then we are going to need something more . . .

More than this

The days and nights in Santa Fe were intense. Late nights sitting in the sports bar or in the dormitories common room, discussing what we had learned, but up early for the lectures, Max positioned in the front row, his pencil moving at full speed across his notepad as Rodriguez spoke. Alex in the back row, sitting with his legs stretched out over the row of seats in front of him but still listening carefully. Esther, Zamya and Madeleine in a row in the middle of the lecture theatre. Next to Madeleine, but separated by two seats, was Antônio. I sat next to him. Rupert sat on his own, but near enough that he could hear both what we were talking about and everything Rodriguez said.

At the last lecture of the first week of the four-week programme, Professor Rodriguez summarized what we had learned about statistics. Both the good and the bad. She reminded us that the relationship between cancer and smoking can be seen as a story of the successful use of statistics. Despite a conspiracy – involving tobacco manufacturers, doctors, scientists and politicians – to prevent research on the link between the two, the truth eventually emerged. Smoking causes cancers.

But Fisher's mistakes, she told us, were emblematic of an arrogance which arose out of the scientific progress of the twentieth century. At first, it appeared that everything could be answered by experiments and observations, but the question of how all these details would go together was often forgotten. In Fisher's case, he had reduced all the information regarding society's problems to a simplified question of genetics, to a dangerous notion of 'feeble-mindedness'. Decades of research on the link between smoking and cancer was reduced in his work to a question of whether or not smokers inhaled.

'It is our job,' Rodriguez told us, 'to think bigger than this. To understand the deeper connections. And this cannot be done by statistical thinking alone.'

The way to find connections was to change our point of view, she said. Instead of seeing the world from above, as if we were all-powerful and all-knowing, we should see it from below. We should realize there is more than one way of thinking about the world.

'There is so much more than this,' she said, her hand pointing a finger at the projection of Fisher on the screen, as if she were telling off a naughty schoolboy, 'and that more is what you are going to learn in the coming weeks.'

The discussion that evening was intense. Everyone seemed to have an opinion on Professor Rodriguez's lectures. Max loved them. He felt she had really broken down the strengths and weaknesses of science going into the new millennium. 'The next twenty-five years are going to be so exciting. We are going to move on from the reductionist approach of just throwing statistics at everything and start really taking complexity into account,' he said.

Antônio also approved. He told us that Rodriguez hadn't yet covered the latest theories, but everything she said about the past was spot on. Madeleine took a more pragmatic view. She used statistics in all her work in biology, and she felt that it would be unwise to throw away everything we knew before establishing what the alternative was. Esther nodded in agreement. 'Rupert explained it to me very clearly the other day,' she said. 'To do mathematical modelling, we just need to keep our heads and be sensible in our analyses.'

Rupert felt emboldened by Esther's praise. 'I think the problem with Rodriguez is that she is big on criticism, but lacking solid answers.' Rupert quoted Nobel prize-winning economist Kenneth Arrow, whose maxim was that a mathematical model was only useful if its results could be explained in words. In Rupert's view, Rodriguez needed to explain how any alternative approach was going to work. Up to now, it had just been empty words and snide insinuations about his hero. He thought it was inappropriate to attack Fisher in a personal way.

'Fisher is a historical figure!' exclaimed Zamya. 'Parker is attacking what his science, and the scientific thinking of that time, represents. We have to do better.'

'Just wait until next week,' said Max, trying to change the subject, 'then we will find out.' Professor Parker was from the Institute of Advanced Study at Princeton, he reminded us. That was where Einstein had worked when he came to the USA. Parker was going to explain a new approach. One that breaks down systems in terms of their interactions.

I could see that Rupert was frustrated by Max's enthusiasm, but he also knew that Parker's Princeton address meant that he would need to be taken seriously. 'Let's just wait and see,' he said.

'Max is right,' Antônio said. 'Parker is going to be amazing. His work deals with dynamical systems and chaos – all the really cool stuff. There are some amazing results there, things we can apply to understand population change, weather systems, economic collapse. Everything!'

'You seem to know so much already, Antônio, maybe you should give the lectures,' joked Madeleine.

Class II: Interactive Thinking

The cycle of life

Let's go back to the start of the twentieth century, to find a new hero. One that we hope won't let us down, like Fisher did.

It is 1902 and Alfred J. Lotka is a final-year undergraduate student in chemistry at the University of Birmingham in the UK. He is an excellent student, but something about his training has left him empty inside. For his current experiment, he carefully mixes together an acid and an alkali. They react with each other, forming water and salt. Patiently, Alfred swirls a glob of undissolved salt in the bottom of the beaker, waiting for something to happen. But nothing does. Nothing ever happens. The experiment is over. The equilibrium is reached. The results are noted, the products weighed, and the next experiment begins.

In the evenings, Alfred, just like Ronald Fisher, reads Charles Darwin. Here he finds complexity and patterns, a cycle of life and an explanation of existence as a never-ending struggle. Where, Alfred wonders, does chemistry fit in? Where is the reaction that turns grass into cows? What is driving the legs of the fox as it chases the rabbit into its warren? What produces the cycles of thought in his mind? If life – chaotic, kaleidoscopic life – really is built from chemical reactions, like his teachers say, then why is chemistry itself so boringly stable?

He reads Herbert Spencer, the nineteenth-century social scientist and philosopher who coined the phrase 'survival of the fittest' to explain Darwin's theory. Spencer writes about how conflict in nature results in 'every species of plant and animal . . . perpetually undergoing a rhythmical variation in number'. And Spencer doesn't stop at biology. He writes about the ceaseless movement of our emotions, our thoughts and our societies. It is these vibrations of life that Alfred wants to work with, not the flat end products of his chemical experiments, not the silent salt at the bottom of the beaker.

When Alfred looks at the calm exterior of his professors and classmates, he is nervous about revealing how much Spencer's words are shaking his own thinking out of balance. His readings have raised questions that he can't answer; questions he can't let go of. He is an immigrant, of Polish background, and has tried hard to fit in among his very English colleagues. He has learned to keep his emotions in check, to talk politely about the effectiveness of the recently arrived batch of Bunsen burners and the plans for the new tearoom.

Eventually, Alfred raises the confidence to ask his favourite tutor why chemistry lacks life. The tutor doesn't know the answer either, nor does he completely understand the young Lotka's question, but he is sympathetic. He tells Lotka about someone who might be able to help: Professor Wilhelm Ostwald in Leipzig. Ostwald has rejected the central notion that molecules are the fundamental building blocks of chemical reactions and has turned instead to the physics of thermodynamics as the explanation of the rich array of patterns that make up biological life. Like Herbert Spencer, Ostwald is looking for a first principles explanation that can thread a link all the way from physics, through biology and on to the dynamics of our own social behaviour.

Alfred Lotka moves to Leipzig for a year of graduate study and follows Ostwald's lectures. While Spencer and Ostwald have the same goal, their approaches are very different: in his books, Spencer evokes complexity with flowery words, while Ostwald emphasizes mathematics and calculations. Alfred sets about learning calculus and differential equations, the mathematical tools that Ostwald claims will reveal the secret.

And gradually, he starts to see a glimpse of how his problem could be solved. What if, instead of doing experiments in the lab, he could do his experiments in his head? This was what Einstein was also doing (unbeknown to Alfred), sitting in a post office in Bern, Switzerland. It was what Fisher would later do in his rooms in Cambridge. Mathematics was the tool that would allow him to conduct thought experiments with rigour and precision.

Lotka leaves Leipzig after a year and moves to the US to find employment, first with the General Chemical Company and later as a scientific editor. In the evening, he continues to work and it is there, sitting at his desk, that he has an idea: to do chemistry that isn't chemistry. He writes down the following chemical reactions:

$$R \rightarrow 2R$$
$$R + F \rightarrow 2F$$
$$F \rightarrow D$$

At first sight, these three reactions were like the ones everyone learned in school. For example, the chemical reaction for hydrogen gas reacting with oxygen gas to form water is

$$2H_2 + O_2 \rightarrow 2H_2O$$

It says that two hydrogen molecules react with one oxygen molecule to form two water molecules. And we can use the same terminology to talk about Alfred Lotka's chemical reactions. The first reaction says that R 'molecule' can spontaneously form two R 'molecules'. The second says that an R 'molecule' reacts with an F 'molecule' to form two F 'molecules'. And the final reaction says that an F 'molecule' becomes a D 'molecule'.

So far, so good. But, as any good chemistry teacher knows, chemical reactions must be 'balanced': the number of atoms must be the same on both sides. The reaction for water formation is balanced: there are four atoms of hydrogen and two atoms of oxygen on the left side of the arrow, and four atoms of hydrogen and two atoms of oxygen on the right side. But Lotka's reactions ignore balance. The first equation has two Rs on the right, but only one on the left. The second has an R and an F on the left, but two Fs on the right. Lotka's model was evidently contradicting a fundamental rule.

But it was here that the insight came. Lotka had realized that by ignoring chemical balance, ignoring stability, he could produce the pattern he was looking for: he could produce the cyclical dynamics of life itself.

Rabbits and foxes

Professor Parker held a piece of white chalk in his hand as he pointed to the large three-section blackboard behind him. In the top-left-hand corner, he had written the same three chemical reactions that Alfred J. Lotka had first described in 1910

$$R \rightarrow 2R$$
$$R + F \rightarrow 2F$$
$$F \rightarrow D$$

Parker explained that we should see the R molecules as rabbits and the F molecules as foxes. The first reaction in Lotka's chemical reaction system, $R \rightarrow 2R$, means 'that rabbits breed like . . . well, rabbits breed like rabbits,' said Parker, smiling at his own joke. If rabbits are left to themselves, unbothered by foxes, one rabbit will soon become two rabbits. The second reaction, $R + F \rightarrow 2F$, says that when foxes eat rabbits, more foxes are created: once a fox has eaten its fill of rabbits, it will start to make baby foxes. The final reaction, $F \rightarrow D$, tells us that foxes eventually die as well.

Parker told us that it is best to see this as an abstract model of female rabbits and foxes, and to assume there are enough male rabbits and foxes hopping around waiting to do their stuff so that the female rabbits can keep reproducing whenever they want to. By making a few (not entirely realistic) assumptions, Lotka's reaction is a reasonable model of how predators, like foxes, affect the population of a prey, like rabbits.

He showed us how to rewrite chemical reactions in a form known as differential equations, which describe change in time. To convert chemistry into maths, he asked us to imagine a field of grass in which the rabbits and the foxes are running around, more or less at random, bumping into each other like molecules in a beaker during a chemistry experiment. Then he asked us to work out what the population of foxes would be when the rabbits are eaten at the same rate as they are born.

'That's easy,' blurted out Rupert, who was sitting next to me. 'It's supply and demand. When the rate at which the rabbits give birth is equal to the rate at which foxes eat them, then their numbers aren't changing. They are at equilibrium.'

'That's right,' said Parker. 'There is an equilibrium at which rabbit numbers don't change.'

Parker had already drawn out the axis of a graph on the board. The x-axis was for rabbits, the y-axis was for foxes. Now he drew a line from left to right across the graph (the horizontal dotted line in figure 4a). 'Every point on this dotted line is an equilibria,' he said. 'The number of foxes eating rabbits is balanced with the birth rate of the rabbits – meaning the rabbit population will neither increase nor decrease.'

Parker went on to draw another line, this time for foxes, vertically from the top of the board to the bottom, which he told us was the equilibria for foxes. It showed the number of rabbits needed for a stable fox population, when a sufficient supply of prey was required to balance out against inevitable fox deaths (the vertical dotted line in figure 4a). He explained that the lines divided the blackboard into four different zones, or quadrants, within each of which the rabbits and foxes had different growth patterns. He started at the bottom-right zone, where there are lots of rabbits but very few foxes. 'Here,' he said, pointing at the board, 'there are not enough foxes to stop the rabbits from reproducing, but those foxes that are around have plenty to eat, so both foxes and rabbits increase in number.'

It was important to notice, Parker said, at any point within this bottom-right quadrant, in which he had just drawn an arrow pointing up and to the right, that both rabbit and fox numbers increase. He then motioned upwards, to the horizontal line which corresponded to rabbit equilibria, and said, 'Once I cross this line, when the fox numbers have increased enough, then the rabbit numbers

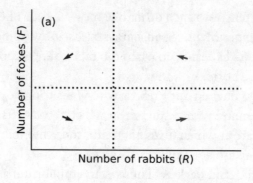

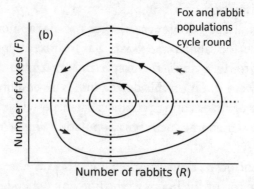

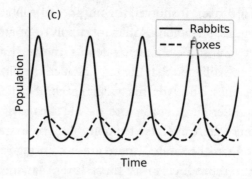

Figure 4: Parker's illustration of Lotka's predator–prey model. (a) The horizontal dotted line shows the equilibrium where the rate at which foxes eat rabbits is the same rate as the rabbits reproduce. The vertical dotted line shows the equilibrium where the rate at which foxes reproduce is the same rate as they die. The arrows show, for each of the four quadrants, whether the foxes and rabbits decrease or increase within that quadrant. (b) Here we include the cycle of the population of rabbits and foxes. (c) The same cycle of rabbits and foxes, but now shown over time.

will decrease.' He now drew an arrow, in the middle of the top-right zone, pointing upwards (to indicate increasing fox numbers) and to the left (to indicate decreasing rabbit numbers). The professor continued around the board, showing, for each of the four zones, what direction the arrows pointed (the arrows in figure 4a).

'Now here comes the true insight,' Parker said. 'If I now follow these arrows around the board, then look . . .'

Parker's chalk now moved on a trajectory that followed the direction of the arrows (the solid lines in figure 4b). Starting from the bottom-right corner where rabbits outnumbered foxes, it went up to the top right, where there were lots of foxes and rabbits. Then, as the rabbits were eaten by the foxes, their numbers decreased and, when he reached the top-left quadrant, the number of foxes also started to decrease. And finally, as he entered the bottom-left section of the board, the rabbits started to increase again as the foxes decreased. As he crossed back, into the bottom-right quadrant, the cycle started again.

'And this is why we never reach stability,' Parker said. 'The interactions between the species take us on an endless cycle.'

Parker explained that, although he was doing this calculation by hand, the same thing happened if we simulated the equations on a computer. Parker put up an overhead slide showing how the cycles around the board (figure 4b) translated into changes in fox and rabbit populations over time (figure 4c).

Rupert and I had been copying down the equations the professor had written on the board next to his graph. At first, Rupert had been muttering small counterarguments to himself, and to me. He was trying to find a hole in Parker's reasoning, convinced that eventually the supply of rabbits should cancel out the demand by foxes and the two populations should stabilize. This was how he had been taught to reason about economic models in Oxford. He thought that the same equilibrium should apply here.

But each time he thought he had found a mistake, Parker would, even before Rupert had time to pose the question, explain why Rupert's potential objection was wrong. Parker had admitted that there were some problems with the original mathematical model proposed by Lotka, but that these problems had been ironed out by other researchers in the decades following Lotka's work. In systems of interactions, cycles were just as common as stability. And these cycles could be found everywhere around us . . .

'What is amazing,' Parker said, looking at Rupert and me, 'is that all cycles . . . the electrical pulses that pass through our brains, the beating of our hearts, the flashing of fireflies in the night, the rapid propagation of turning in a flock of starlings, the spread of an epidemic, the rise and fall of fashions and the boom and bust of our economy . . . these are all patterns which emerge from interactions of individuals: a single brain emerges from billions of separate neurons; the flock emerges from each unique bird; the economy emerges from people buying and selling goods.'

Parker explained that the key to Lotka's method was describing how each individual component of the system affected the other components. In the example on Parker's blackboard, the components were foxes and rabbits. When neuroscientists create a model of brains, the components are the neurons themselves and the chemical and electrical signals they send between them. In modelling insect swarms and bird flocks, the components are the animals. And in modelling our societies or economic systems, the components are us – individual humans.

Parker said that the father of modern economics, Adam Smith, had been wrong, because his stable thinking had convinced him that the market would reach and stay at equilibrium. But Smith's thinking was, Parker said, reductionist. Accounting for our interactions, the way we also behaved like animal herds, showed that human society was anything but stable. We experience the same ups and downs as rabbit and fox populations. We are in constant flux.

There were now only two minutes left of the lecture, and Parker stood there, silent, letting the effect of his words sink in. He looked down, quietly considering what he wanted to say. When the right words did come, they were almost in a whisper.

'You see, this stuff is almost like magic,' he said, one hand gesturing faintly towards the blackboard behind him, now filled with drawings and equations. 'It is a way of seeing those things that others can't see. If you know this stuff. If you know how to see the interactions, the dynamics of causation, if you like, then you know how to see the truth. The world is not stable. The reductionist theories that came before Lotka – that still permeate so many of our notions about science – they just blind us. The stuff on this blackboard behind me is the way of seeing. It is the way of finding how interactions produce patterns that are more than the sum of their parts.'

Rupert looked at him intensely. It was clear he wanted to say something, to tell Parker that he was going too far by talking about magic. But he also saw the calculations covering the board, the ultimate mark of authority and rigour. These results had been thought through.

To me it seemed clear. Rupert was beaten. Back at Oxford, he was being taught a form of mathematics and statistics that dealt only with stable equilibria. Parker was suggesting something different. Something that better captured my experience of the world. Something I wanted to better understand.

Social chemistry

Let's now look at our own worlds through the lens of chemical reactions. Imagine that the world consists of two types of people: Ys who smile and Xs who don't smile. When a grumpy X meets a smiley Y, then X also starts to smile. We can write this, precisely as Lotka did, as

$$X + Y \rightarrow 2Y$$

When a smiler meets a non-smiler, we have two smilers. The reaction isn't a real chemical reaction. There may well be some form of chemical reaction going on in our brains when we smile, but it isn't this that we are trying to capture here. Instead, we are capturing a simplified form of personal 'chemistry' that goes on between two individuals when one is smiling. We are describing a social reaction.

Let's try another one. Imagine there are cops (C) and robbers (R). If a cop meets a robber, she arrests him. That is

$$C + R \rightarrow C + A$$

The cop is still a cop, but the robber has now been arrested (A). In this example, the state of the robber has changed, but the state of the cop stays the same. And another. Imagine a person who is trying to move a sofa into your living room from outside the house. We use the letter O to represent a sofa that is outside, and L a sofa that is in a living room. We let P be a person who is trying to move a sofa. The following reaction

$$P + O \rightarrow P + O$$

says that if one person tries to move a sofa, he won't succeed, and the sofa will still be outside the house. But if he gets his friend, then

$$2P + O \rightarrow 2P + L$$

Two people together ($2P$) can move the sofa.

We can apply the same rule to our smilers. Imagine that just one person in a group is smiling, laughing or having a good time. This is not enough to persuade others to do the same. After all, one smiling person might just be a deluded madman, laughing maniacally at their own nonsensical joke. But if two people smile, then it is more likely that you will smile too. Two people are less likely to be crazy than one. And so the smiling reaction becomes

$$X + 2Y \rightarrow 3Y$$

It takes two smilers to infect one non-smiler to make three smilers. We will call this the 'it takes two' rule, and we will look at the dynamics it creates in more detail later on.

Think of a few of these interactions in your own life. It can be anything from how gossip spreads between friends: one person who knows the gossip tells another who doesn't, making it two of them who know the gossip. Or how you and your colleagues complete a task together at work. The job goes a lot more quickly when two people work in partnership. These interactions can be as trivial as how you might transform a pile of dirty dishes into sparkling clean dishes. Or they can involve your internal state of mind: perhaps the very act of doing the dishes moves you from a state of lethargy to having a small sense of achievement. The interactive view of the world is written in terms of chemical reactions: this can be social chemistry with others or individual chemistry when we think about our own state of mind or the state of the world around us.

We don't need to be too concerned about whether or not the reactions we write down are always true in all situations. Sometimes a single person can manage to move a sofa into their living room; often the cop doesn't catch the robber or two of your friends can laugh at a joke even though it isn't funny, but that isn't the point of this exercise. The starting point for this way of thinking, an interactive approach, is to look at our own life in terms of how we change the world and how the world changes us. We should see ourselves as part of a social reaction: we act in a certain way, and that changes the way others around us act. Equally, the actions of others shape how we act and think.

Interactive thinking is different from the statistical view of the forest of people. It is more individual. More personal. More related to our everyday experiences. It relies less on data and more on thinking through the consequences of our actions. As we shall see, it captures how people make the same choices and decisions as their friends, how conformity spreads through groups, and how our moods swing up and down. But it is no less scientific than the forms of stable, statistical thinking we have seen up to now. In fact, it often provides much more comprehensive answers to some of the most important questions in our society.

The social epidemic

One of the most important applications of Lotka's chemical reaction method has been in modelling the spread of a virus during an epidemic. When someone who isn't infected by a virus but is susceptible (and not immune) is in contact with another person who already has the virus and is infective, she can also become infected. In Lotka's language of chemical reactions this is

$$S + I \rightarrow 2I$$

A susceptible S plus an infective I becomes two infectives.

At the start of an epidemic, almost everyone is susceptible, since very few people have had the virus. If an infective person comes into contact with one new person every second day, then by the end of the second day she will have infected one person and two people will have the disease. After four days, the two infected people both contact one person each and there will be $2 \times 2 = 4$ infectives. After six days, there will be $2 \times 2 \times 2 = 8$, and on day eight there will be $2 \times 2 \times 2 \times 2 = 16$. The infection doubles every second day. By day twenty, there will be $2 \times 2 \times 2 \times 2 \times 2 \times 2 \times 2 \times 2 \times 2 \times 2 = 1{,}024$ infectives.

A shorthand for multiplications of this form is to write, for example, $2^3 = 2 \times 2 \times 2$, which we call 2 to the power of 3. The number of times we multiply, three in this case, is known as the exponent. In our example above, the exponent is the number of days since the first infection divided by two. So, for example, the number of cases on day six is $2^{6/2} = 2^3 = 2 \times 2 \times 2 = 8$, and the number of cases on day twenty is $2^{20/2} = 2^{10} = 2 \times 2 \times 2 \times 2 \times 2 \times 2 \times 2 \times 2 \times 2 \times 2 = 1{,}024$. This type of growth, where the exponent for growth is proportional to the number of days since the first infection, is known as exponential growth.

Exponential growth occurs very rapidly. By day forty, the number of infectives is $2^{20} = 1,048,576$ (2 multiplied twenty times). On day sixty, it is $2^{30} = 1,073,741,824$, which is slightly more than one billion. For a real virus, like Covid-19, there is a time period between when a person is infected with a virus and when they can infect other people – so the virus does not double every second day, but the growth is still exponential: the number of infected people multiplies over time. And this repeated multiplication causes the case numbers to become very big, very quickly. Before we know it, the virus is everywhere.

Initially, exponential growth leads to a large number of infections, but after some time the infectives start to recover, following the reaction

$$I \rightarrow R$$

Infectives (I) enter the recovered state (R). As a result, when infectives meet other people, the probability that they are susceptible decreases, since many of the people they will meet have already recovered. As a result, the growth of infectives reaches a peak and starts to fall. This is shown in figure 5a, for a mathematical model, known as the SIR model (S for Susceptible, I for Infectives and R for Recovered), which is based on the two chemical reactions listed above. The initial growth is rapid, like when we kept multiplying by two, but as the number of recovered people increases, there are less and less people to infect, and the disease fades away.

As more people are infected, the rate at which the disease spreads will decrease, because many of the infectives will come into contact with already recovered individuals, i.e. people who are immune. Let's assume, for example, that a person is infected for a week before recovering. When they are the only infective person, they will on average infect 3.5 other people (one person every two days). Once half of the population is either infective or recovered, then an infective person will infect only 1.75 people because only $3.5/2 = 1.75$ of the contacts will be with susceptible individuals (the other half will be with recovered or already infected individuals who don't spread the disease).

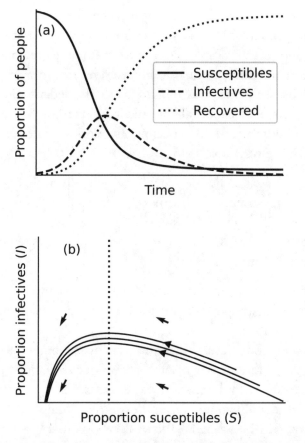

Figure 5: SIR model, (a) How a typical epidemic develops over time. (b) The same epidemic, but showing now how the relative number of susceptibles (x-axis) and infectives (y-axis) are related.

Once only 1 in 3.5 of the population is susceptible, then an infective individual will produce, on average, 3.5 / 3.5 = 1 new case. It is here we say that *herd immunity* has been reached: each infection will lead to less than one future infection. Figure 5b shows the same sort of diagram that Parker drew on the board for the predator–prey model, but now for the SIR model. This type of plot is known as a phase plane. Instead of plotting the proportion of susceptibles and infectives through time, we plot them against each other. The arrow then indicates the direction of time. The dotted line indicates the level of infection at which herd immunity is reached: this is an equilibrium at which the growth of infection goes from positive to negative.

Unlike the stable, class I thinking we met in the previous section, the class II thinking we carry out here does not rely primarily on data (we use only one piece of data in the above discussion: 3.5 the average number of people an infected individual comes into contact with). It builds instead on reasoning. By working through the consequences, we were able to

1. Track how the disease grows exponentially initially.
2. Give an estimate of how many people will be infected eventually.
3. Get an idea of what level of vaccine coverage is required for herd immunity.

All these important conclusions are derived from only a few clearly stated assumptions in the form of chemical reactions for our social interactions.

Epidemic models are an essential part of responding to epidemics, but they are also very useful in everyday situations that have nothing to do with diseases. Culture, ideas, jokes, behaviour and fashion are all contagious. We use terms like 'viral video' (one that spreads quickly between TikTok or Facebook users) without pausing to think about just how powerful this analogy is. The power of writing down our social interactions as chemical reactions is that it gives us the freedom to develop and test our own ideas about the social world, while still thinking rigorously and mathematically.

Suki always wants to know about the latest trends, to be the first one to share a funny meme with her friends or to know what line of clothing is most fashionable. But she often finds that, no matter how much time she spends online, she is seldom ahead of the curve. The others seem to have already seen the funny dog videos she shares or heard about the latest Off-White drop about the same time as she does, even though they are much less interested than she is. Why is it that she can't get ahead of the game?

Let's imagine a situation where the number of people who have seen a meme has doubled every hour up until Suki's friend Sofie hears it, ten hours after it first appeared. Sofie isn't that quick when it comes to following social media. So, ideally, Suki would like to know it well before her. Let's say five hours ahead.

To understand why it is difficult for Suki to be five hours ahead of Sofie, we need to step backwards through time. If 100,000 people (including Sofie) had seen a meme before the end of hour ten, then half that, 50,000 people, had heard it before the end of hour nine; 25,000 before the end of hour eight; 12,500 before the end of hour seven; 6,250 before the end of hour six; and only 3,125, or roughly 3 per cent, had heard it before the end of hour five. In other words, the time it took for the first 3 per cent of people to see a meme is the same as the time it took for the other 97 per cent of people to find out. If Suki wants to be ahead of the curve, she has to work extremely hard to get into that 3 per cent.

In general, if we divide a typical epidemic curve, that in figure 5a for example, into a start, a middle and an end, we see that the majority of 'infections' occur in the middle. There is a fast-growing start, a middle where many people are infected, and a tail where the last people are infected. During any particular social epidemic – when, for example, you share news or a meme online – you are much more likely to be in the middle section than at either of the extremes. When you hear about something, it is likely that this is exactly the point that almost everyone else hears about it too.

For many social behaviours, it isn't just the adoption of a fad or a news cycle that is contagious, but also the way in which we recover. When we get a cold, flu or Covid-19, the best thing to do is to go home, rest and not spread the virus. Spending time with people who have already been ill doesn't help us recover more quickly (even if their sympathy might help us feel a bit better). Recovery is independent between individuals. This is reflected in the chemical reaction $I \rightarrow R$: no other individual is required in the recovery reaction.

For fashion and news trends it is different. For example, when Richard stops talking about the TV show *Game of Thrones* because he notices Antony has lost interest, we can think of this in terms of social recovery. Richard recovers from his earlier obsession with the show more quickly, because he starts meeting others who have also recovered. In this case, the chemical reaction is

$$I + R \rightarrow 2R$$

When people infected with a fad meet a recovered individual, they themselves recover more quickly.

The $I + R \rightarrow 2R$ recovery reaction means that social epidemics are different from virus epidemics. In particular, it makes 'vaccination' much more effective, because those people in the recovered state now cause infected people to recover more rapidly. For disease epidemics, herd immunity is given by the vertical line (the dotted line in figure 5b) which we need to reach before cases decrease. When social recovery is possible, the 'herd immunity' line (the dotted line in figure 6a) is rotated. As a result, epidemics which start with no 'recovered' individuals are almost as wide-reaching as in the non-social recovery model (figure 6b), but when, for example, 30 per cent of the population is already socially recovered, the infection dies out rapidly (as in figure 6c).

A common public relations trick used by companies when viral 'bad news' appears about them is to place a follow-on story in the media that is similar to the original bad news but with a more positive slant. This new story not only presents the company's point of view, it also exploits social recovery. When those infective people who have heard the original 'bad news' story tell it to those who heard the 'positive slant' version, they feel that they are sharing yesterday's news. The story the infectives heard now appears less original than this new version, and they stop telling it. The trick to controlling the news agenda is to focus less on the infectives – those spreading the 'bad news' – and more on shifting susceptibles to recovered, thus dampening interest in the original story.

Sofie has been working on a project to spread information about the importance of getting vaccinated for Covid-19. Sometimes she

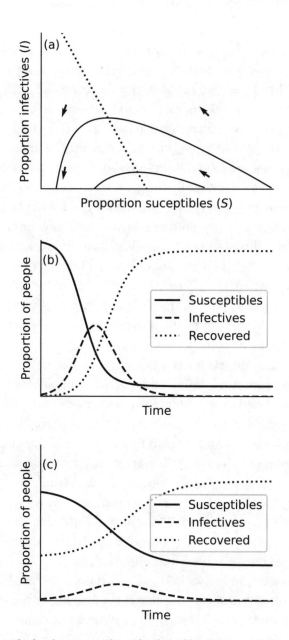

Figure 6: (a) The herd immunity line (the dotted line) is rotated when social recovery is possible, in comparison to figure 5b. (b) When the initial number of recovered is low, the infection spreads through most of the population. (c) When the initial number of recovered is 30 per cent, the infection dies out quickly.

can be frustrated by all the false information that spreads about vaccinations and feels she needs to argue against each piece of fake news. Then she remembers the importance of recovery – not just recovery from Covid-19 itself, but also the social recovery we need from the bogus science that surrounds it. If Sofie can educate people, provide accurate information, then not only will they be critical when they hear disinformation, they will also inoculate others. Instead of focusing on changing minds and worrying about those who simply won't be swayed, she instead socially vaccinates the people around them.

Social contagion is very often a force for good. After catastrophes, such as tsunamis or devastating storms, donations towards rescue and recovery follow an epidemic curve: we denote money because we see others doing so. Our social interactions – when we start a round of applause, how we laugh together or when an audience lets out a shocked collective 'oh' at a Dave Chappelle joke – are contagious. We are continually sensing out each other's approval in order to find the right thing to do.

Social contagion and recovery can stretch over much longer time scales than a rumour, a news cycle or a round of audience applause. Up until the 1960s, Irish setters were not a particularly sought-after breed of dog. The American Kennel Club registered two or three thousand puppies every year. But then the breed's popularity began to grow exponentially. By 1967, there were ten thousand Irish setter registrations, and at the peak of their popularity, in 1973, there were sixty thousand. After that the drop started, and it went quickly: from fifty-five thousand in 1975 to thirty thousand in 1977 and back at ten thousand by 1980. By the 1990s, Irish setters were even less popular than they had been in the 1960s.

Similar boom and busts in owning certain dog breeds have been experienced with Dobermann pinschers peaking in the late seventies, chow chows, which peaked in 1987, and rottweilers peaking in the mid-nineties. On average, it takes a breed around fourteen years to rise from obscurity to a popularity high and a further thirteen to fall back down to a low again. In some cases, the boom is started by films, such as the re-release of Disney's *One Hundred and One*

Dalmations in 1985, which resulted in a 700 per cent increase in registrations before the end of the decade, followed by a precipitous drop in popularity by the mid-nineties.

This same form of contagion and social recovery has long-term effects on our lives. Participants in the Framingham Heart Study – a multigenerational study of the lifestyle and health of tens of thousands of inhabitants of the city in Massachusetts, in the USA – were found to be twice as likely to drink excessively if they have a friend who drinks excessively, and more likely to abstain from alcohol if a friend (randomly selected by the researchers) abstained. They were almost two and a half times more likely to smoke tobacco and three times as likely to smoke marijuana if a randomly selected friend smoked these substances. Similar results are found for the tendency to be obese and time spent sleeping. They even hold for divorce: having a divorced friend increases the probability of a person also being divorced. Divorce is possibly the most extreme form of social 'recovery': we are more likely to end the most important relationship in our lives, just because our friends have done the same.

When I draw parallels between buying a particular breed of dog, sharing news online, laughing together at a comedy show and a person's decision to leave their partner, it might appear as if we are trivializing some big life changes. The distinction to make, though, is between studying psychological mechanisms – which of course differ greatly between sharing things online and the disintegration of a marriage – and similarities in the dynamics of these systems. The role of our friends in the stability of our romantic relationships is more complex and long term than their role in determining which pictures we share on social media. Similarly, the way we choose a dog breed is very different from how we might fall into alcohol dependence. But the underlying social reaction is the same. If we want to model how rottweilers became popular in the US or how a group of high-school students start drinking together, then we use the same chemical reaction and we find the same rise and fall dynamics for both.

With this knowledge comes a deeper responsibility for each other. When you engage in a negative behaviour towards someone else it affects not only that person but those close to them, because your negative behaviour spreads.

Our ten friends in London are always starting new projects. Last summer, Becky had them growing vegetables at an allotment; during the winter, Antony tried to get them to play five-a-side football once a week; and shortly after that, Jennifer started to run a book club. In all these activities, everyone initially pulls in the same direction and participates. But after a while interest wanes: the friends get bored of constantly weeding the vegetable beds and the members of the book club lose interest after a particularly boring book. It seems that interest in each activity dies out as quickly as it started. Such cycles of engagement are natural and unavoidable. Instead of feeling frustrated, or blaming themselves for not insisting that the group see things through, Becky, Antony and Jennifer should look back on what they have achieved as part of a natural cycle. It lies in the nature of social infection and recovery that our interest ebbs and flows.

Sometimes, we can have a false belief that stable things are better: that the 'true' news is a long-term trend, and not all the viral stories; that, in order to be fulfilled, we should see through all the projects that we start; that it is our average happiness over time that is important; that we should hold true to certain values throughout our lives . . . But stability is not the outcome we should expect from social interactions. It is the highs – the times when we really did achieve something together or when we are enjoying ourselves the most – that really matter. Our own interactions, just like those between the fox and the rabbit, pull us backwards and forwards, from collective elation to gloomy despondency, from following one news story and forgetting about another, from one social activity to the next, between different ideas and beliefs, towards one life goal and away from what we thought we wanted.

Letting ourselves drift in the comings and goings of our interactions isn't irrational. What is irrational is a belief that things are better when they are stable.

More than the sum of its parts

As we leave the lecture theatre on the Wednesday of the second week, after Parker's lecture on models of epidemics, Madeleine, the Australian biologist, grabs my shoulder from behind, turns me to face her, looks me straight in the eye and says, 'We need to talk.'

She guides me to one of the tables set up outside for our afternoon coffee break and starts to explain. The lecture has had a profound effect on her, she tells me. This was just what she had been looking for. Parker's description of interactions was key. Couldn't I see it too? It was exactly what she needed for her ants. 'Ants aren't stable,' she says. 'They're always doing something different. Switching between tasks, like cleaning and feeding their brood, trying to find new food sources, building sections of their nests . . .'

Madeleine had been collecting data on how ants foraged for food for the last two years, but she hadn't been able to see the bigger picture. This was what Parker had provided. But the problem was that she couldn't do the maths herself. 'That's where you come in,' she says, smiling and still looking directly at me. My job is to help her describe ants in terms of Lotka's chemical reactions.

This was exactly the challenge I had been looking for too. I take out my notebook, while Madeleine continues to talk about 'her' ants. She always referred to them like this, as if they were her children. She tells me about the pheromones they laid on the way to food. She tells me about their cycles of activity: sometimes they were running everywhere; other times they all lay still inside their nest. And all the time, while she is talking, I write: one reaction describing the ants collecting food at the feeder, one reaction describing resting ants, another to describe ants searching for food. Madeleine keeps correcting the diagrams I draw, explaining that something is too simplified or that what I have written down isn't as important as I thought it was.

The other students go back into the theatre to listen to the next lecturer, but we continue to sit at the table. Madeleine emphasizes that when only one ant found food and left pheromone for the others, this would often evaporate before the others could follow the trail. 'If we think of finding food like catching a virus – it would take at least two ants to infect another,' she says. As soon as she says it, I realize: the chemical reaction for ants recruiting each other to food must be something like

$$L + 2F \rightarrow 3F$$

It takes two ants which have found food (F) to recruit one ant that is looking (L) for food. Two ants who have found the food convert one ant who is looking for it. This is the 'it takes two' reaction.

We add one further reaction

$$F \rightarrow R$$

which says that, over time, ants which have collected food eventually retire (R). This is similar to recovery in an epidemic model, and we assume that ants retire without consulting the others. 'They drift off and find something else to do,' explains Madeleine.

By now, Madeline has taken my piece of paper from me and is sketching reactions herself. I find another sheet and start to write down equations describing the dynamics of the above reactions – differential equations, which describe changes over time, in terms of the rate at which the reactions occur. First, I looked at the reaction rates when the colony is small in size or the ants are looking for food over a very wide area. In this case, the ants meet infrequently and the rate of recruitment to the food is low. So, even if some of the ants find the food initially, knowledge of the food can die out because ants retire before they spread the word about the food. This is shown in figure 7a, where the arrows point downwards to where there are no longer any ants that have found the food (all those who found it earlier have retired).

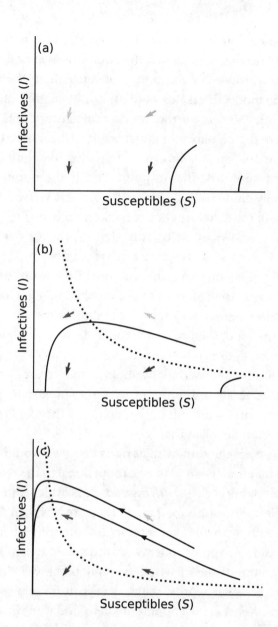

Figure 7: The 'it takes two' model of ant trails. An infected ant is one that has found food and a susceptible ant is one that hasn't. (a) When interaction rates are low, the infective ants fail to recruit many susceptible ants to food. (b) When interaction rates are moderate, the infective ants fail to recruit many susceptible ants to food if the initial level of discovery is low, but succeed if it is above the 'herd immunity' line. (c) The infective ants almost always succeed in recruiting to food.

Then I looked at what happens when the rate of interaction increases. To do this, I calculated the equilibrium at which the rate of recruitment was equal to the rate of retirement. For the standard epidemic model this is the 'herd immunity' equilibrium, where the rate of infection equals the rate of the recovery. Unlike an epidemic, where the equilibrium is a straight vertical line (the dotted line in figure 5b), for our model of the ants, the equilibrium is a curved line (the dotted line in figure 7b). If the number of ants which initially find the food starts below this curved line, then these ants will retire before they recruit enough ants to spread the word (see the arrows to the bottom right below the dotted line in figure 7b). However, if a significant proportion of ants find the food initially, then they are able to spread the word and recruit nearly all the ants to the food. Finally, I looked to see what happens when the ants interact very frequently. In this case, if only a few ants found the food, then almost all of them would eventually find out where it is (figure 7c).

'It is just like Parker was talking about in the lecture!' I exclaimed. 'The ants are literally more than the sum of their parts. We can't just sum them up to find out how much food they will collect; it is much more complex than that.'

If the ants were the sum of their parts, then we would expect the proportion that find the food to be proportional to the rate at which they interact with each other. Instead, ants which interact infrequently collect almost no food at all: they are less than the sum of their parts. Ants with a higher interaction rate sometimes collect a lot of food and sometimes collect very little: they are either more or less than the sum of their parts, depending on how their search for food began. In other words, their success in foraging for food is determined by chance: if enough of them find it in the first place, then they will all find it (more than the sum of their parts); otherwise, very few of them will find it (less than the sum of their parts). Very large colonies are no longer subject to randomness; they will always be successful in getting most of the ants to food.

'Small colonies always fail. Large colonies always succeed. And medium-sized colonies' success depends on how many find the food to start with,' I summarized.

Madeleine was very excited. 'I can test this!' she said. 'I can do an experiment. This is going to be so much fun.'

I wasn't entirely sure how she would do it, so she explained. By splitting the ants into different-sized colonies she could manipulate how much they interacted. The model predicted that small colonies would be completely unable to establish a pheromone trail to food, while large colonies would be disproportionally successful.

Now I got exactly what she meant. And I realized something else. For medium-sized colonies, everything depended on how the search for food started. 'Is there a way you can help the ants get going,' I asked, 'so we can test whether there are two stable outcomes, one where they collect lots of food, another where they get almost none?'

'Yes! Of course,' she answered, explaining that she could do an experiment where she placed some of the ants at the feeder to start with, so they got some help finding the food initially. The model's prediction was that ants given a head start in this way would always make a trail to the food. Those without a head start would not usually build a trail.

'Ants are just like us humans,' Madeleine said. 'I imagine if a large enough group gets going with the work, then they will all get involved. That's how we work and I'm sure the ants are the same. I am going to test this as soon as I get back to Sydney!'

Start a fitness craze

The way in which ants rapidly find food, after a small group first discover it, is an example of a tipping point. In our own society, tipping points can describe situations where a trend that has been growing very slowly suddenly (and for no apparent reason) takes off and becomes extremely widespread. Take, for example, men in their twenties growing beards. Up until around 2012, growing a beard wasn't particularly popular among young men in the UK. That all changed quite quickly, and a few years later beards of all shapes and sizes were to be seen everywhere. We would say that late 2012/early 2013 was a tipping point for facial hair.

There are some similarities between tipping points and social contagion and recovery. Both involve one group of people influencing another group to join them. But tipping points are different in that they involve two stable states: one where almost no one engages in a behaviour or a fashion, and another where lots of people engage. It is the dotted equilibrium line in figure 7b which allows for two stable states: one with no ants finding food; the other with ants swarming over the food. Or, to take the example above, one with very few beards; another with lots of beards.

The model of ants in the previous section reveals that multiple stable states, separated by tipping points, result from the 'it takes two' infection reaction

$$S + 2I \rightarrow 3I$$

We can see this type of social reaction in humans, just as we see it in ants. For example, in our group of friends, Antony was the first to grow a beard, but initially John and Charlie were reluctant to copy him. The change came when Richard decided to grow a goatee.

Suddenly John took notice: they are one *S* (John) and two *I* (Antony and Richard), and John is convinced. He starts to sport a closely trimmed three-day stubble. It isn't long until Charlie follows suit.

A tipping point, or critical mass as it is also referred to, is sometimes characterized by saying, for example, 'He grew a beard because everyone else grew a beard.' In fact, the conditions for a tipping point are more like 'He grew a beard because a couple of his friends had them, and then a fourth friend copied him and one of the friends, and it spread from there.' A tipping point does not require people to be aware of how many people in the population as a whole are growing beards, wearing pink shirts, watching a particular Netflix series or even getting tattoos or committing petty crimes. Rather, a person simply has to have a sense that a significant number of people around them are doing these things. Local transmission of ideas or behaviour are sufficient to create a tipping point.

Imagine that Jennifer wants to get in better shape physically. She knows, after finding out about the twelve years of extra life to be gained from a healthy lifestyle, that she should exercise regularly and drink in moderation, but the problem is that her friends are just like she is – they don't exercise more than once a week and many of their social activities revolve around heavy drinking at the weekends. She would like to make the change but doesn't feel able to do it alone. How can she get herself and her friends to change their ways?

It is in situations like this that understanding the difference between the one-to-one transmission reaction we see in the spread of a virus and the 'it takes two' reaction above is essential. Now and again, one of Jennifer's friends does suggest that they go for a jog together. They put on their trainers and plod round the park. But it never becomes a regular occurrence, because a few days later another friend suggests you skip the jog and go to the pub instead. They all sit there laughing at their futile attempts to get fit.

To overcome the barrier, Jennifer needs to think about her friends like a difficult-to-move sofa: getting someone on her side to help would make things an awful lot easier. She remembers when Antony had a beard and John didn't consider one for himself until

Richard started sporting his goatee. Jennifer isn't as close to Nia as she is to Sofie, but she knows that Nia is likely to follow up on the commitments she makes. So she suggests that the two of them go out for a light jog, twice a week. Nia has a tendency to work late, so Jennifer texts her during the day they have agreed to go running, reinforcing how well it went last time and telling her that she is looking forward to seeing her in the park. Soon, they have formed a bond and Nia starts texting Jennifer to remind her as well. Jennifer doesn't worry about the others . . . yet.

Then, when the two of them have got into a routine, Jennifer realizes that it is time to involve Sofie. She invites her out for a run and asks Nia to do the same. They don't always go jogging all three together, but now Jennifer's focus lies in getting Sofie engaged. Nia is already hooked, and is happy to message Sofie and invite her out when Jennifer is busy.

For a group of ten friends, three isn't yet enough to take them over the tipping point. And not all the ten friends will be persuaded to go jogging. If pushed to do exercise, Aisha and Suki would rather do aerobics; Antony, Becky, Charlie and John prefer playing football; and Richard (if really forced) prefers to go to the gym. So Jennifer creates a group chat with all the ten friends called 'exercise twice a week'. She, together with Nia and Sofie, share pictures of their runs. She arranges an evening when they all play football together and she books Aisha and Suki in for a free aerobics class. With the help of the two she has already converted, it is becoming easier to convert the others. She still has to make an effort, because the stable state for this group is to be resting in the pub (especially after playing football), but she knows that, once she gets to the tipping point, it will be worth it.

Now comes the beauty of the tipping point. Once the group is over the threshold of five (Suki starts aerobics and Charlie starts posting pictures from five-a-side), the feedback keeps them there. The peer pressure within the group is now to stay healthy. If Jennifer starts to drift back to her previous, unhealthy behaviour, then her friends will remind her that it is time to go running or to take a drop-in aerobics class. Even the most reluctant members of the group start posting pictures in the group chat.

The important point to remember is that the effect is not proportional to the effort put in. Initially, Jennifer needs to work really hard to convince her friends, but once she is over the threshold, then very little effort is needed to keep all of them there. Again, this is different to the one-to-one infection model, where the effect is directly proportional to the number of contacts between individuals. In a standard epidemic model, the initial work needed to get healthy behaviour started is less than required in the 'it takes two' reaction, but once a behaviour is established, *more* work is needed to keep everyone going in the one-to-one model than the 'it takes two' model. Once everyone is part of Jennifer's fitness programme, she can relax (a bit); if she starts to feel demotivated, she will quickly recover because she is now surrounded by fitness-crazy friends.

In our own interactions, the lesson is that if we want to make a change for the better, we need to increase the intensity of our interactions. It isn't enough to try something once, we need to build momentum within a group. Once we have that momentum, when we have reached the stable state where everyone is involved, then it will be easier to keep going.

Often, in group settings, at work and school, we find ourselves trapped in a vicious cycle where it seems that everyone is negative all the time, and any attempt to be positive is met with further negativity. Simply trying to make one or two positive comments per day might lead to you hearing a few positive comments said back to you, but it won't be enough to change the culture, because your low-intensity positivity will be drowned out by the high-intensity negativity. But this doesn't mean that it is impossible to change the group dynamic. You need instead to sit down together and agree to try to change. Maybe not everyone will adopt a positive attitude directly after this meeting, but once enough of them do, then the tipping point will do the rest of the work for us: those people who resisted will eventually be persuaded by the positivity of those who embraced a change.

A third law

The dream is to find a system. To find a way of capturing the essence of what we see. One set of reactions. One list of simple rules.

It is 1920, ten years since Alfred J. Lotka published his cycling reactions, and he still feels his life to be unfulfilled. There were certainly some interesting challenges to be found at work – analysing data from experiments, editing journal articles and examining patents – and his colleagues appreciate his analytical skills. But these were, to Lotka, trivialities. They didn't lead him deeper, to the understanding he craved so much.

The response of fellow academics to his original article was underwhelming. It had received almost no attention, and no one had followed up on it. Lotka had felt some small encouragement after reading an essay published during the First World War by a lieutenant-colonel in the British army, Sir Ronald Ross, who was studying the spread of malaria. Ross had recognized what he described as two ways of thinking about epidemics: a posteriori and a priori methods. He wrote that 'in the former we commence with observed statistics, endeavour to fit analytical laws to them, and so work backwards to the underlying cause [. . .]' This was the statistical approach which Cambridge statisticians, like Fisher, were developing.

Of more interest to Lotka was what Ross called the a priori approach. Ross argued that in modelling a malaria epidemic – where the aim is to capture a complex interaction between mosquitos, the virus itself and the humans who get ill – it is an a priori approach that is needed. Of this approach he wrote, 'we assume a knowledge of the causes, construct our differential equations on that supposition, follow up the logical consequences, and finally test the calculated results by comparing them with the observed statistics'. We 'assume knowledge' that mosquitos spread the malaria virus randomly between human hosts, 'follow up the logical consequences' using an epidemic model to plot how the disease will spread, and 'test the calculated results' by comparing the model to epidemic curves. Ross had sought help from mathematician Hilda Hudson in solving the models he proposed, and together they used them to explain the rise and fall of epidemics. Hudson even found the same cyclical behaviour seen in Lotka's model, and she and Ross then used it to explain why epidemics have second and third waves.

Lotka realized that it was the a priori method which he was trying to create, but not just for epidemics. Ross called it a theory of happenings, which hinted at something deeper. But Lotka didn't feel that 'happenings' captured the true depth. He thought back to his studies with Ostwald in Leipzig. There the focus was on the second law of thermodynamics, a law for physical systems, which states that as energy is transformed, some energy is always wasted, in the form of heat. The second law is the reason that all real-world chemical reactions eventually reach equilibrium: the reactions between molecules in a chemist's beaker will in time become balanced and the molecules themselves evenly dispersed. It seemed to Lotka that the second law did not hold for living systems, which were characterized by perpetual creation of new plants and animals. It was precisely the second law that he had ignored in order to create his own oscillating reactions. He wondered if he could find a new law of thermodynamics that could be applied to biological systems, to social systems and even to our consciousness. A third law.

He started by making a few small improvements to his 1910 model and published it in a leading US journal, the *Proceedings of the National Academy of Sciences*. The editor, Raymond Pearl, was so pleased with Lotka's submission that he asked for a further article. Encouraged, Lotka started by taking all the thoughts that were whirling through his head and writing them down. In 1920s USA, after the war, after the Spanish flu pandemic, he was witnessing a society re-creating itself, coming back to life. He wrote that his fellow humans were 'accelerating the circulation of matter through the lifecycle, both by "enlarging the wheel", and by causing it to "spin faster".' Are humans driving some as yet unknown physical quantity toward a maximum? he speculated. 'This is now made to appear probable,' Lotka wrote, 'and it is found that the physical quantity in question is of the dimensions of power, or energy per unit time.'

After this new paper, Raymond Pearl was even more convinced that Lotka was on to something. Pearl invited Lotka to Johns Hopkins University, found him a two-year-long research fellowship, and told him to let the ideas flow, to find the source of the power that was driving the human cycle, to write down all those thoughts that had been going round in his head for the last twenty years.

In terms of ambition and scope, Lotka did not disappoint. The resulting book, *Elements of Physical Biology*, quoted from the Bible, H. G. Wells, Lewis Carroll and Wordsworth. It described the growth of sunflower seeds and rats, the spread of bacteria colonies, increases in the population of the United States, worms feeding on cadavers, epidemics of malaria, forest ecosystems, parasites and hosts, food webs, male life expectancy, practices for breeding and slaughtering cattle, and the ups and downs of economic imports and exports. In each case, the solution, according to Lotka, came back to one idea: to write down the chemical reactions and study the interactions.

Today, the approach taken in those early days by Lotka, Ross and Hudson is widely applied across all types of biological systems. We have already seen how it is used to model the spread of diseases. It is also used to model cancerous tumours and to find new ways to stop them growing. It is used to describe ecosystems and how they will respond to climate change. It is used to understand the pattern of stripes on a zebra's coat and the development of animal embryos. It is even used to model the biochemical reactions that form the basis of life itself.

Cellular automata

During the Friday lecture of the second week in Santa Fe, Parker had been showing us examples of how Lotka's approach could be applied in everything from neuroscience to the modelling of weather. At the end of the lecture, he introduced one of his colleagues: a man with shoulder-length hair, cowboy boots, Levi jeans, a checked shirt and a prominent belt buckle. I was so fascinated by his hippy-cowboy appearance that I didn't catch his surname when Parker first introduced him. But the cowboy told us we could call him Chris.

'I didn't know we were going to the rodeo at the weekend,' whispered Rupert to Esther and me.

Esther laughed. I was less amused. I had hoped that Rupert would have admitted by now that he was actually learning a lot from Parker's lectures, but he continued to make sarcastic comments, many of which Esther seemed to find highly amusing. It was this which bothered me the most. Surely she, having studied with Parker during her masters project, should understand just how amazing his theories were?

They both quietened down, and when Chris started speaking the lecture theatre was completely silent. He told us that in the afternoon he would be running a computer lab on cellular automata. These were the types of mathematical model on which Stephen Wolfram based his theoretical investigations. Wolfram didn't invent cellular automata models, Chris told us, but he specified what are known as elementary cellular automata, the most fundamental model of this sort. The computer lab was voluntary, and Chris said he understood if we would rather have the afternoon off, but we were welcome to join him.

'See you later,' said Rupert. 'I have proper equations to solve. I'm not going to waste my time with computer games.'

Esther turned to me and asked, 'Are you coming?' I didn't have to think about the answer. An afternoon off would be nice, but this was something I had to be part of.

Once inside the computer lab – a basement room containing a variety of machines of different shapes and sizes running a range of operating systems – Chris told us that the aims of Lotka and the aims of Wolfram, seven decades later, were not all that different. Both of them were puzzled by the second law of thermodynamics, which tells us that in physical systems things get more and more disorganized and random over time. The pressure stabilizes throughout a gas, salt dissolves throughout a glass of water, a fire eventually burns out to leave a pile of stable ashes and the heat from the fire diffuses and cools through the air.

Why then, both Wolfram and Lotka wondered, is the world full of periodic patterns like predator–prey cycles, and even more complex patterns, of which life itself is perhaps the most complex of them all?

While Lotka had focused on chemical reactions, where all the individual molecules floated around freely in a gas or a liquid, Wolfram realized the importance of local interactions. 'In epidemic models or predator–prey models,' Chris explained, 'we assume that individual animals (or humans) are equally likely to come in contact with any other individual. But in a cellular automata model, interactions are set up on a fixed grid of cells. This is much more like real life, where we meet the same people and interact with them over and over again.'

To understand cellular automata, Chris asked us to consider the following string of ones and zeros (known as a binary string)

$$110010000111000011011111001$$

We refer to each 1 or 0 in a binary string as a bit, in the same way as we refer to the numbers between 0 and 9 as digits. So, just like the decimal number 458 is made up of three digits (4, 5 and 8), the binary string 010 is made up of three bits (0, 1 and 0). The binary string above thus consists of twenty-seven bits.

Elementary cellular automata are rules which tell us how to convert one string of bits into another string. For example, consider the following two rules which we will apply to the bits in the string above

1. If a 0 has a 1 to its immediate left, it becomes a 1;
    otherwise, it remains a 0.
2. A 1 always remains a 1.

If we apply that rule to the binary string, we get a new string

$$111011000111100011111111101$$

Notice that all the 0s with a left-neighbouring 1 in the original string have now become 1s in the new string. Applying this rule one more time, we get

$$111111100111110011111111111$$

And again

$$111111110111111011111111111$$

And finally

$$111111111111111111111111111$$

All the bits in the string are now 1. Chris explained that, for any initial binary string, all the 0 bits to the right of the first 1 will eventually become 1s. Step by step, the 1s propagate through the 0s, changing them all to 1s, like dominos falling down in a domino rally. A stable arrangement of 1s.

Chris showed us another string

$$010001101111010101011010$$

and asked us to apply the following three rules to each bit in the string:

1. If both of a bit's neighbours (to the left and right) are 0, then it becomes a 0.
2. If both of a bit's neighbours are 1, then it becomes a 1.
3. If a bit's neighbours have different values, it should remain the same number as before.

For example, if the three bits are 010, so the middle bit is 1 and both its neighbours are 0, then it should (according to rule 1) change to a 0, so the string becomes 000. Applying rules 1–3 to the longer string above produces the following change

01000110111101010011010

↓

00000111111110100111100

↓

00000111111111011111100

↓

00000111111111111111100

All the isolated 0s or 1s are replaced by their majority neighbours (we assume that the bits form a loop, so the 1 on the end of the string has neighbour 0 to its left and adopts its right neighbour as the 0 at the start of the string and thus becomes a 0).

He told us that we could, if we wanted, think of the 1s and 0s as people with political opinions. The 0s might be Democrats and the 1s Republicans. Each person has two neighbours: if both their neighbours disagree with them, then they change their mind; otherwise, they keep the same opinion. As a result, clusters of Democrats and clusters of Republicans form.

Chris explained that this pattern was stable: if we applied the rule to it again, we would get the same output.

00000111111111111111100

↓

00000111111111111111100

No matter how many times we applied the rule, the string would never change, and no matter what string we started with, stability would be achieved – as long as people stay in the same place, don't hear new opinions and don't change their view of the world. It is like politics in the USA, he joked, a string of Republican 1s in the middle of the country and Democrat 0s on the coasts.

'Now,' he challenged us, 'I want you to simulate this rule on the computer. These can make some far-out patterns, if you know how. But we are still getting started, so let's see if you can create something that looks like a chequerboard using these rules.'

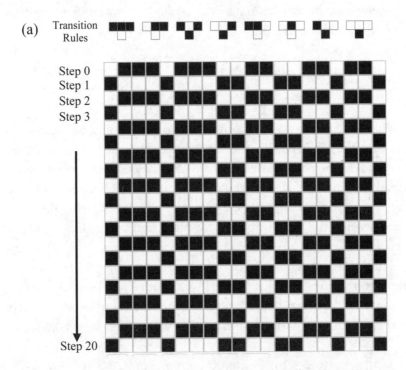

Figure 8: A simple periodic pattern in an elementary cellular automaton. The top of the figure shows the rules for the cellular automaton: how the three cells on the row above determine the outcome on the row below. These rules are those given in the text: black is a 1 and white is a 0. In this case, when the middle cell of the three is white then the output cell below is black, and vice versa. This creates the alternating line pattern.

There weren't enough computers in the lab for everyone to work on their own, so I sat down with Esther. She took firm control of the keyboard, angling it towards her. 'This isn't going to take long,' she said, and started typing code into the computer terminal's window.

Esther quickly realized that if we made a rule which simply turned all the 1s into 0s and all the 0s into 1s, it would endlessly loop. She implemented this rule as an array of black (representing 1) and white (representing 0) cells and very soon her screen displayed the pattern shown in figure 8. Chris leaned over our shoulders and nodded approvingly. 'Groovy lines,' he said.

Esther paused for a minute. She had realized, she said, that all elementary cellular automata rules could be written in a way that looked something like this

111	110	101	100	011	010	001	000
0	0	1	1	0	0	1	1

Here, the top row shows how the two neighbours of the central bit, and the bit itself, on the row above determine what that bit becomes in the row below. The rule set above is the one that generated lines: in every case, if the central bit is a 0 it becomes a 1, and vice versa. In the computer a 1 is black and a 0 is white – the rules are shown at the top of figure 8 for black and white cells – and the simulation creates alternating black and white lines.

Finding a rule set for the chequerboard was more difficult to create than the lines. And just when I was about to make a suggestion, before I could get more than a couple of words out of my mouth, Esther exclaimed, 'Got it!', and quickly wrote out a new rule set

111	110	101	100	011	010	001	000
1	0	1	1	0	0	1	0

As she typed it into the computer, she explained more. 'If we keep the all-black areas black and the all-white areas white, then the chequerboard will eat in from the sides,' she said.

The rules she proposed were mostly the same as the one that had produced alternating patterns. But, where all three of the cells in the line above were black, respectively white, then for her new rules, the middle cell remained black, respectively white. It wasn't a perfect chequerboard, it didn't always match at the points where the black areas collided with the white, but it was exactly what Chris had asked for (figure 9).

He looked very pleased. 'Groovy,' he said again. 'Now for the challenge! I want you to create a set of rules that makes a chaotic pattern, a set of rules that makes complex patterns, and one that makes beautiful patterns. I want you to use your imagination. And next week, you can show them to me.'

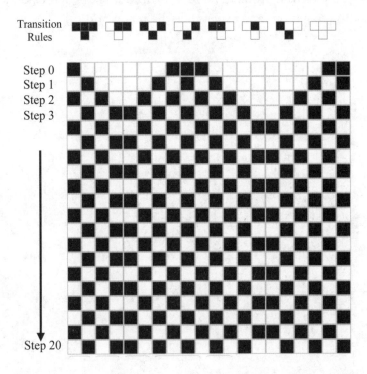

Figure 9: A chequerboard pattern in an elementary cellular automaton. The top of the figure shows the rules for the cellular automaton: how the three cells on the row above determine the outcome on the row below. These are the rules Esther explains in the text.

Esther and I sat for a while longer, trying different ideas out. But every cellular automata rule we tried created either a stable black screen, a stable white screen or alternating patterns. It was getting late on the Friday, and I was no longer feeling particularly inspired.

Chris came up to the computer we were sitting at. He was on his way home from the lab.

'Take a break, guys,' he said, switching off the computer's monitor. 'You need to get out of here and soak in the vibes downtown. There's this bar I know, called El Farol. Go there this weekend. I promise you won't be disappointed.'

Esther and I looked at each other. Chris was right. We weren't going to solve the problem he set us without taking a break first. It was time to get out and experience the real world.

The art of a good argument

Charlie and Aisha have been a couple for five years and have two small children. They have a good relationship, but with both of them working, enjoying a busy social life, looking after the kids and trying to keep on top of the housework, tempers can fray. Sometimes, directly after they have had an argument, Charlie wishes that he had a recording of the conversation. He would like to rewind the tape, go step by step through the argument, and show Aisha why it was her (and not him) who had caused tempers to rise. To explain that while he was just trying to stick to the facts, she had made it personal.

Most of us have felt this way. If only we could show *them* how *they* had driven the discussion in a wrong-headed direction. How *they* had lost their composure. It was *their* fault. *They* started it. Or so *we* think.

We definitely should *not* record our intimate conversations. Secretly recording conversations is a recipe for broken relationships and divorce. Even alluding to such a recording is enough to re-enflame the argument. Charlie telling Aisha, 'If only you could hear yourself' is itself an error, not a way of helping her see (what are from Charlie's point of view) the error of her ways.

But . . . it is worth thinking through (on our own, when we're much calmer) about what those recordings might contain and how *you* sound in them.

To start this process, let's create a model of a heated discussion between Aisha and Charlie. It is important in this context to understand that Aisha and Charlie love each other. Very much so. They care about their relationship, and they respect each other at a deeper level, but just at this very moment they happen to be having a disagreement. Quite a nasty disagreement.

We are going to represent their behaviour as a binary string, a sequence of 1s and 0s. For example, the sequence below might represent an escalating frustration on the part of Aisha

Aisha: 0 0 0 0 0 0 0 1 1 1 0 0 1 1 0 0 0 1 1 0 0 1 1 1 1 1 1

Each bit represents one sentence she has said: 0 indicates a calm response, while 1 is Aisha raising her voice. As the conversation goes on, the 0s become less frequent, and the 1s become more frequent. Clearly, Aisha is getting more and more upset.

This is, though, the sound of just one partner talking. What is it that is driving the change from calm to angry? To find out, let's now look at Charlie. Again, a 0 is a calm response and a 1 is aggression

Charlie: 0 0 0 0 0 0 0 0 0 1 1 0 0 0 1 1 1 0 1 1 1 0 1 1 1 1 1

He is also getting annoyed. As the sequence goes from left to right, Charlie is switching from 0s to 1s. It takes two to have an argument.

Now let's place the two strings next to each other

Aisha: 0 0 0 0 0 0 0 1 1 1 0 0 1 1 0 0 0 1 1 0 0 1 1 1 1 1 1
Charlie: 0 0 0 0 0 0 0 0 0 1 1 0 0 0 1 1 1 0 1 1 1 0 1 1 1 1 1

Since time goes from left to right, we can see that Aisha produced a shouting 1 before Charlie did.

'Well, that's it then,' Charlie might proclaim. 'Look at who got angry first. Aisha, you started it . . . I was just responding to provocation.'

The 'they started it' accusation is something we learn from a very young age. For many of us, it is a moral law: if you are the one who started it, then you are the one to blame. But this view is wrong, and it is dangerous. I know because, in this case, the sequences of 1s and 0s I have produced are not from a real argument, they are outputs of a mathematical model.

Let me first explain the model, and then we will return to this conclusion: that 'they started it' is the wrong way to see an argument. Consider the following transition rule:

$$50 \text{ per cent:} \quad 1$$
$$0 \rightarrow 1$$

It should be read as follows: 'If Aisha has just shouted at Charlie, but Charlie hasn't just shouted at Aisha, there is a 50 per cent probability that Charlie shouts at Aisha.' The 1 in the top line indicates that Aisha shouts; the 0 in the bottom line indicates that Charlie is not shouting, and the $\rightarrow$ 1 indicates that Charlie starts shouting. The 50 per cent defines the transition probability.

The rule above is *probabilistic*. We can think of Charlie as tossing a coin when Aisha shouts at him. If the coin lands heads, he starts shouting back; if it lands tails, he doesn't shout. This is an important difference from the elementary cellular automata that Chris showed us in Santa Fe. Those were *deterministic*: for any given input, they always do the same thing. In applications of cellular automata, like our current application to arguments, a probabilistic approach is more realistic. Humans don't always do the same thing in the same situation; we are fundamentally unpredictable. The probabilistic rule captures some of our unpredictability.

So far, I have defined only one rule for a conversation. In general, the rules depend on what Aisha and Charlie did most recently in the conversation. So, for example, if neither of them is shouting, then there is a lower probability (we will say 10 per cent) that Charlie starts shouting, and we write

$$10 \text{ per cent:} \quad 0$$
$$0 \rightarrow 1$$

This probability isn't zero. Aisha and Charlie are having an intense discussion, so Charlie may very well lose his temper, but the probability of this is much lower than the probability of shouting when the other person is shouting. Following a similar logic, we can make transition rules for Charlie for all possible states of shouting:

'Neither shouting'	'Aisha shouting'	'Charlie shouting'	'Both shouting'
10 per cent: 0	50 per cent: 1	70 per cent: 0	95 per cent: 1
$0 \to 1$	$0 \to 1$	$1 \to 1$	$1 \to 1$

The first two rules are those we discussed above: if neither Aisha or Charlie is shouting, there is a 10 per cent chance that Charlie starts shouting; and if Aisha is shouting, there is a 50 per cent chance that Charlie starts shouting. To these we add two more rules: if Charlie is already shouting (but Aisha isn't), there is a 70 per cent chance that Charlie continues shouting; if both are shouting, there is a 95 per cent chance that Charlie continues shouting and a 5 per cent chance that they spontaneously stop.

We have defined how Charlie reacts to every possible situation. Let's now do the same for Aisha

'Neither shouting	'Aisha shouting'	'Charlie shouting'	'Both shouting'
10 per cent: $0 \to 0$	70 per cent: $1 \to 1$	50 per cent: $0 \to 0$	95 per cent: $1 \to 1$
0	0	1	1

Notice that Aisha and Charlie react in exactly the same way to each other. They are as bad- (and as good-) tempered as each other and respond to similar cues.

We now have a complete model for two people arguing, and it was this model which I used to create the sequences for the feud between Charlie and Aisha. Let's look at this sequence again, now in the context of the rules that generated it.

Aisha: 0 0 0 0 0 0 1 1 1 0 0 1 1 0 0 0 1 1 0 0 1 1 1 1 1 1
Charlie: 0 0 0 0 0 0 0 0 1 1 0 0 0 1 1 1 0 1 1 1 0 1 1 1 1 1

Initially, pairs of 0s are usually followed by further 0s, because of the 'neither shouting' rule, which has a low probability of resulting in shouting. When, by chance (there is always a 10 per cent probability), Aisha starts to shout, now there is both a greater chance of Charlie starting to shout, and Charlie continuing to shout. For a while, there is some flipping backwards and forwards between who shouts, until they get stuck in a sequence of both shouting.

The sequence above is just one possible simulated argument, arising from our probabilistic cellular automata model. By running the model lots of times we start to get an idea of the types of outcomes it can produce. Sometimes the transition to both people shouting will happen more quickly, for example

> Aisha: 0 0 0 0 1 1 1 0 1 1 1 1 1 1 1 0 1 1 0 1 1 1 1 1 1
> Charlie: 0 0 0 1 1 1 1 1 1 1 1 1 1 1 1 1 1 0 1 1 1 1 1 1 0

Sometimes it will take longer time to start, but then continue

> Aisha: 0 0 0 0 0 0 0 1 0 0 0 0 0 1 1 1 1 1 1 1 1 1 1 1 1
> Charlie: 0 0 0 0 0 0 0 0 0 0 0 1 1 1 1 1 1 1 1 1 1 1 1 1 1

Sometimes there will be a mix of bickering 0s and 1s

> Aisha: 0 0 0 0 0 0 0 0 0 1 1 0 1 1 1 0 1 1 0 0 0 0 1 1 1 0
> Charlie: 0 0 0 0 0 0 1 1 1 0 0 1 1 1 1 1 0 0 0 0 0 0 0 0 1 1

Sometimes both of them will be a bit grumpy, but no argument will occur

> Aisha: 0 1 0 0 0 0 0 0 0 0 1 1 1 0 1 0 0 0 0 0 0 0 0 0 0 0
> Charlie: 0 0 0 0 0 0 0 0 0 0 0 0 0 0 0 0 1 1 1 0 0 0 1 0 0 0 0

And in some cases the shouting will start up, but then quickly die down again

> Aisha: 0 1 1 1 1 1 1 1 1 1 1 1 0 0 0 0 1 0 0 0 1 0 0 0 0
> Charlie: 0 0 0 1 1 1 1 1 1 1 1 0 1 0 0 0 0 0 1 0 0 0 0 0 0

115

Each of these arguments comes from the same set of underlying rules for the pair's interactions, the rules I set up above, but each of the outcomes is very different.

The 'they started it' rule does not provide valuable insight into the simulated arguments shown above. Sometimes Aisha starts it; sometimes Charlie starts it. Sometimes Aisha shouts more than Charlie; sometimes it is the other way round. But we know (because we set up the rules) that the way Charlie and Aisha respond to each other is the same. They both trigger each other in the same way. The specific argument Charlie and Aisha happen to have just had is irrelevant, it is the rules of interaction that created it which are fundamental.

What we do, by setting up a probabilistic cellular automata model, is look at the consequences of a certain set of inter-actions. This then gives us a starting point for looking at how Aisha and Charlie might improve how they talk to each other. Instead of trying to record all the things that Charlie doesn't like Aisha doing, Charlie should instead look at how to adjust his response. For example, imagine if Charlie changed his own rules to the following

'Neither shouting'	'Aisha shouting'	'Charlie shouting'	'Both shouting'
10 per cent: 0	10 per cent: 1	10 per cent: 0	95 per cent: 1
$0 \rightarrow 1$	$0 \rightarrow 1$	$1 \rightarrow 1$	$1 \rightarrow 1$

Under these rules, Charlie tries never to raise his voice in response to Aisha and makes sure that if he does (mistakenly) raise his voice (which happens 10 per cent of the time), he tries to immediately stop. But Charlie also realizes that if both of them are shouting, then it is going to be hard to stop, so the same rule as before applies in this case.

Here is an example discussion simulated under these new rules

Aisha: 0 0 1 0 0 0 0 1 1 0 0 0 0 0 0 0 1 1 1 1 1 0 0 0 0 0
Charlie: 0 0 0 0 0 0 1 0 0 0 0 0 0 0 0 1 0 0 0 0 0 0 0 0 0 0

In this sequence, Charlie makes two mistakes, which set off Aisha shouting (Aisha is still following the same rules as before and is likely to shout when shouted at), but since Charlie was able to stop himself and not respond, a larger argument is avoided. This strategy won't avoid all arguments. In this simulation, for example

Aisha: 0 0 0 0 0 1 1 1 1 1 0 0 0 0 0 0 0 1 1 1 1 1 1 0 1 1
Charlie: 0 0 0 0 0 0 0 0 0 1 0 0 0 0 0 0 1 1 1 1 1 1 1 1 1

Aisha and Charlie both started to shout simultaneously, and then it was difficult to stop. But, in general, the discussions arising from this new model, where Charlie tries to avoid shouting in response to Aisha, involve a lot less bickering.

Ultimately, the only person that we can truly change is ourselves. But if you change how you respond to others, the underlying rules of your interactions, then you will also change the outcomes of those interactions. Charlie shouted less, so Aisha also shouted less, not because Aisha changed their underlying rules, but because Charlie gave her less negativity to respond to. Charlie made the change, and things got better for both of them.

In long-term relationships couples can work on improving together, as long as both parties want to change the rules of their interaction. Integrative Behavioural Couple Therapy (IBCT) is a form of couple therapy that adopts an approach that focuses on doing just that. Two IBCT pioneers, Andrew Christensen and Brian Doss, describe relationships as defined by the interactions partners have with one another – and say that these interactions are themselves determined by the characteristics each partner brings to each situation. This is exactly our way of thinking with the probabilistic cellular automata: the focus shifts from the outcomes (e.g. the disagreements and issues within a relationship) to the inputs, to the rules that govern how we interact.

The role of the therapist in IBCT is to first identify the differences between how the partners deal with conflict. Let's say that in our example, Charlie is withdrawn and doesn't want to talk about emotions, while Aisha needs confirmation and closeness. Shouting on Charlie's part is usually in the form of a frustrated comment about Aisha being too demanding. When Aisha loses her temper, it is usually to accuse Charlie of not listening to her properly. When Aisha complains that Charlie doesn't listen, he feels that this confirms that Aisha is demanding. He pushes her further away. As a result, Aisha feels more isolated than ever.

Both of them are in the wrong, but it only takes one of them to make the first step and break the cycle of arguments. Either Charlie needs to understand that Aisha makes demands because she cares about their relationship, or Aisha needs to see that shouting at Charlie won't make him listen. Once one of them makes the initial change, over time, both of them will notice improvements. And hopefully the other one will also come to see how their previous rules of interaction were causing the arguments.

In trying to change our own rules for the better, we need to think very honestly about our own behaviours. You might not be a person who shouts a lot, but you might tend to make sarcastic comments or sigh in response to solutions suggested by others. It could also be a lack of responsiveness on your part, that you don't acknowledge what the other person says, or deliberately give them the silent treatment. It could lie in your body language; perhaps you are nonchalant, raise your eyebrows or don't look at the other person when they need eye contact. The problem might lie in the way you put forward an argument: maybe you change the subject a lot or you try to make your own opinions sound rational, while suggesting the other person is overly emotional. Or maybe it is you who says a lot of illogical things, reducing the possibility of a reasoned discussion. There are lots of ways in which you might be difficult to deal with.

The key to improving our interactions lies in identifying and talking about the underlying rules. This is very different from the 'replay the tape' approach, which focuses only on specific outcomes and, in its worst form, tries to place the blame on the other person. The best way to improve communication with a person you care about is to do it together. You should find a way to discuss the rules of interaction together. Talk openly about the responses that trigger you and ask about the things that trigger your partner.

Changing your behaviour will bring improvements for both you and the person you care about. A small change to your own rules of interaction can make a massive change for everyone around you.

Top down, bottom up

In one way, Alfred Lotka failed in his grand project.

He failed, in that he never found a third law of thermodynamics for biology. His 1922 book was a catalogue of how to model different phenomena using chemical reactions, but in its five hundred pages it never managed to land on a single, universal insight. Lotka believed that the third law would come from looking at how natural selection acted on the different kinetic reactions. Some chemical interactions, he imagined, produced more 'power' than others, and it was these that survived and reproduced. But he failed to give a convincing definition of what the 'power' of an interaction is. There was no basis for Lotka's theory in biology, and when, decades later, the structure of DNA was uncovered, it was difficult to reconcile the true building blocks of biology, where competition is between individual genes for survival, with the kinetics Lotka described, in terms of interactions between chemicals, species and populations.

Unlike Fisher, who was bitter when rewarded with recognition, Lotka was modest about the little he received. He continued to work on his ideas in the evenings, while throwing himself fully into his work at the Metropolitan Life Insurance Company. There he produced new methods for measuring demographic changes, predicting life expectancy and assigning insurance premiums. He led the development of actuarial science and was made President of the American Statistical Association in 1942. Ultimately, it was his professionalism, rather than any supposed genius, that his peers came to appreciate.

Looking back, one hundred years on, even if there is no grand unifying 'third law', there is another way to see Lotka's way of thinking – interactive and cyclic – as a success. Today, scientists use a variety of ways to talk about class I and class II thinking. Class I

120

thinking is sometimes referred to as top-down. It starts with a theory and then looks at how well that theory explains data: does smoking explain cancer? Does life expectancy explain happiness? Class II thinking is more bottom-up. It starts with observations of how we think the world is – foxes eat rabbits, couples sometimes argue, it takes two people to start a health craze, we influence each other's political opinions – and generalizes these observations to a set of rules. We then derive the consequences to create a theory. Using this approach, we don't start with a cloud of data points from surveys – like we do when applying statistical thinking to health or happiness. Instead, we start by trying to understand the essence of a system: how it works, what are the key components, how do they fit together and when do they fail? From there we make predictions – predator–prey cycles, bouts of shouting, tipping points in beard growing and exercising, political polarization. It is after we have made these predictions that we test them on data from the real world.

I have found that people vary a lot in how receptive they are to these ways of reasoning. Many people feel that class I, statistical thinking – where we start by plotting data and measuring health and happiness metrics, or the occurrence of cancer cases in people who do and don't smoke – is somehow more objective, and therefore better. It is certainly true that we cannot understand the world without data and, like the younger Fisher taught us, some ways of measuring the world are better than others. But the lesson we learned from the older Fisher is that we can't just stare blindly at the numbers. We always bring our own subjectivity to any question; in deciding which data to plot, and which data to ignore.

This is the reason we also need class II, interactive thinking. We need to be able to start from our own understanding and work forward and upward using logical reasoning. This is often my preferred way of working on both scientific problems and personal challenges. Once I feel I understand a problem based on what I already know, I make concrete predictions and test these by collecting and looking at data.

Neither class I nor class II is always right or always wrong. We need to think in both of these ways.

Is that it then? Can we alternate between the bottom-up and top-down approaches to solve any problem that happens to face us? Or, if we can't solve everything, does applying these two methods allow us to approach problems in the best possible way?

Partly, yes. These methods are very useful, but before we can use them with confidence, a rather difficult problem stands in our way . . .

Class III: Chaotic Thinking

Always knowing the next step

Margaret hated making mistakes.

Most of the other students in her mathematics class, all of them young men, carefully memorized every line of the proofs presented on the blackboard. But Margaret had, very soon after the course started, concluded that they were wasting their time. She knew this because she noticed that Professor Florence Long – her teacher at Earlham College in Indiana during the second half of the 1950s – never relied on memory. The professor derived each line of a proof from the previous one, as if seeing it for the first time, carefully illustrating how mathematical reasoning led inevitably to logical conclusions.

This was why Professor Long never made a mistake, thought Margaret. Results reproduced from memory are always fallible. They can't be relied upon because the parts don't necessarily need to fit together. Memorized steps don't support each other. But if each step follows on from another in a logical manner, then errors become impossible. She joked with the boys that it was possible to be lazy (by avoiding memorizing the proof) *and* to never make a mistake. This was what she loved about mathematics – when she knew the underlying logic, she knew she was in control.

Professor Long invited Margaret and her other students to her house for cucumber sandwiches. The professor made them feel at home, and appreciate spending time together, talking and learning. Professor Long was the person Margaret wanted to be. Warm and welcoming. Brilliant and inspiring. Concise and accurate. And never wrong.

Margaret could do anything she set her mind to. As a teenager, she took a job as a tour operator at an abandoned copper mine. Under her stewardship, the enterprise grew from a handful of family visits per week to a thousand visitors per day during the summer. At sixteen, Margaret found herself first organizing a team of tour guides, then opening a copper jewellery shop and finally, by the time she was at university, running the entire copper-mine enterprise (with over a hundred summer employees) herself. The owner just left her to it, preferring to spend his time driving around aimlessly in his collection of luxury cars. Margaret herself took no more in salary than the other employees and so had to hold down jobs waiting tables in the evenings and operating telephone switchboards, all while studying full time (and coming top of her class).

Margaret wanted to pursue a PhD in mathematics. But after she married in 1958, changing her name from Margaret Heafield to Margaret Hamilton, her husband was accepted to law school at Harvard. In 1959, she agreed (somewhat reluctantly) to put her dreams of further study on hold and move to Boston, to get a job to support her husband's studies and their recently arrived baby daughter.

On her first day at her new place of work, Margaret Hamilton thought again about Professor Long. She could see that her new boss, Professor Edward Lorenz, also a mathematician but now working at the metrology department at MIT in Boston, had the same enthusiasm as her former tutor. He opened his office door and showed her his pride and joy, his Librascope General Purpose LGP-30 computing device. This device, he told her, could make any calculation.

Professor Lorenz told Margaret that he believed the machine would allow researchers to predict the weather. For her, though, it was even more than that. The machine was simply one of the most amazing things she had ever set her eyes upon. She had seen calculating devices before, but this was on a different level. It encompassed an ideal: the possibility of limitless logical computation. All she had to do was feed into it paper with punched holes which coded the instructions and it would follow every logical step without fail. Just like Professor Long when she is teaching mathematical proofs, Hamilton thought.

And now the LGP-30 was hers to make use of. None of Lorenz's graduate students had the faintest idea how it worked. Lorenz told her what he knew, handed her the instruction manual and let her get on with it.

She set about learning. During the working day, Hamilton programmed weather predictions, and then, in the evenings, she visited the computer halls at MIT. There she would mix with the self-proclaimed hackers, again all men. They were not used to dealing with women as equals. For them, 'girls were just to go out with'. And, at first, they made the same sexist jokes irrespective of whether she was there or not, seeing her as 'one of the boys'. But Hamilton made clear that she was to be accepted on her own terms: both as a programmer and as a young mother. She took her daughter to the evening lab sessions and coded with her sitting on her knee. She showed the other men that she could be logical and structured in her approach to work and remain a genuine, considerate person in her dealings with other people. And, gradually, the mood in the lab softened. The hackers even took turns to play with her daughter, seeing if she could break their software by entering random inputs.

It was immediately clear to all that Margaret Hamilton had talent as a programmer, but she got her chance to really prove herself when all the men were sent to another town for an obligatory advanced programming course. She was unable to travel because of family commitments, so she stayed in Boston, but she soon realized that this meant she would have access to more computing power while they were away. The plan was that once the programmers had taken the course, they would be able to solve some of the more difficult tasks that had come up at the MIT computing labs. But when the men came back, after two weeks away, Hamilton had not only learned much more than them during her time alone but had completed most of the tasks they had been expected to solve, using the skills learned on the course, upon their return!

While others memorized the instruction book, Hamilton thought in terms of problem solving. Just as she had learned mathematics by understanding how to derive one line of reasoning at a time before attempting to reproduce complex proofs, she now focused on the underlying logic of programming. This was why she could grasp new methods quickly: for Margaret Hamilton each new technique built on the last. The others were often trying to show off, writing intricate code that no one else understood. But a ramshackle bridge built from nailed-together bits of wood, without a proper plan, might get you across the river, but it isn't stable in the long term. Margaret Hamilton took a different approach. She took the logical rigour of Professor Long and used it to solve the problems posed by Professor Lorenz. She was training herself to build code like an engineer: a software engineer.

Nudge

John sees an important part of his role at work as nudging people back on course. The right word of encouragement at the right moment. A word of warning about the direction a project is taking. A well-timed hand on the shoulder. A precisely formulated email. Or even a quiet drink after work. John has perfected a set of tools for handling different situations. And his boss has noticed, giving him increasing responsibility in the projects the company is carrying out.

Becky sees John's approach as cynical. 'You can't treat people like they are little remote-controlled boats that can be steered around,' she says.

But John defends the principles behind his approach. He is rational. What he does is logical. And, a lot of the time, it works.

The image John has in his mind is that people are more like basketballs than boats. His work colleagues bounce around a fair bit, but they have a natural resting point. At that point, John thinks, they are productive and balanced. Whenever they bounce off in the wrong direction, he just needs to get them back to the stable point.

John's thinking is illustrated in figure 10. His aim is to get the basketball to the bottom of the valley, the point marked as 'target'. To do so, he gives it a little push or a nudge in the right direction. Basketballs bounce around a lot, so while his initial nudge takes the ball in the right direction, the second bounce takes it past the target. This is illustrated in figure 10(a): on bounce 2 the ball overshoots. Since the target lies in a valley, though, on bounce 3 the ball rolls back down in the direction he wants it to go and comes to rest at the target. His little nudge has taken the ball to where it should be.

If at any time we notice that a basketball has become lodged at the wrong point, we can just pick it up, give it a bounce, and the chances are it will end up at the bottom of the valley. If it doesn't get

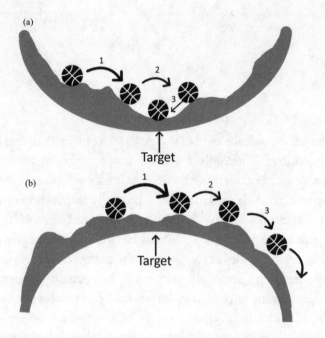

Figure 10: What happens when we bounce a basketball on (a) a stable landscape and (b) an unstable landscape.

there the first time, because it gets stuck in one of the uneven crevices on the side of the hill, another nudge should send it to where we want it to go.

Stability is a property of the structure of the landscape, rather than arising from continual nudges. This is why John's strategy works. He thinks he knows what his colleagues want to achieve; he just has to give them a nudge to get them back to their best.

John is applying interactive thinking in order to make his working world stable. We have already seen several examples of similar approaches. Jennifer started a fitness craze, tipping her friends from one stable state (couch potatoes) to another (fitness-focused friends). When Aisha and Charlie were arguing, they flipped backwards and forwards from calm to shouting. By thinking about how they reacted to each other, they found a stable, argument-free way of interacting.

We also saw how class II thinking dealt with periodic cycles.

Instead of the foxes reaching a stable population, they deplete the rabbit population, which in turn leads the foxes to reduce in number. These cycles are part of societal change: in everything from dog-breed popularity and baby names to economic booms and busts.

If the world around us was made up only of stable states and cycles, we might be able to bring it fully under our control. Think, for example, of US society in the 1950s, the world in which Margaret Hamilton grew up. At this time, engineering was shaping the Midwestern United States. The neat lines of suburban houses, the mass production of family cars, the whirring of time-saving devices in the kitchen, the regular beating of the washing-machine drum and the rotation of the electric gramophone.

Each of these technologies arises from tightly controlled regularities. The gramophone designer keeps the record spinning at regular rotations. The car producer dampens the vibrations caused by potholes in the road. The radio manufacturer amplifies the broadcaster's signal. In each case, the 1950s engineer controls the world to make it more predictable and stable.

It is this characterization of 1950s America that forms the stage in the film *The Truman Show*. Jim Carey's character, Truman Burback, believes he is living a happy post-war family life, blissfully unaware that he is living in a carefully constructed TV set. Under the illusion that he is the hero in his own story, he believes that it is his actions that provide his family with the stability and security of an all-American life.

The problem with *The Truman Show*'s staged view of 1950s America is that the real world doesn't always conform to suburban stability or gramophone-rotating periodicity. The TV production company in *The Truman Show* can't maintain the false narrative of Truman Burback's life for ever. As the lead character bounces out of the confines of his 1950s world, the fake reality becomes more difficult for them to maintain. Chaos ensues.

The same applies to John's basketball view of work. It isn't just Becky who has seen how he manipulates those around him. John's colleagues start to notice, subtly at first, that he is starting to use his method in his own self-interest, to make himself look good or to get everyone to work around his plan. As a result, the landscape around him changes. His colleagues stop listening to him. The stable valley of cooperation becomes an unstable hilltop. The ball is finely balanced on the top of the hill and any nudge will send it flying in an unpredictable direction (Figure 10b). John loses control.

As we explore a new way of thinking, we address a new question: how does chaos arise and what should we do to deal with it?

El Farol

I stood with Alex, the Austrian chemist, at the bar. It was crowded in a way that made me feel at home, like the pubs in England. It was a place where no one would notice me; where I could stand and observe; where I could take in the movement of the people around me without being overwhelmed.

'I saw you with Esther,' Alex said, smiling, 'sitting so close together. Her programming and you trying to look like you knew what you were doing.

'I know what you are trying to do,' he went on, 'but you are going about it all wrong. There is a much simpler way to get there.'

'What's that?' I asked, still not quite clear what Alex was talking about.

'Look around you,' he replied. 'It's right there in front of you. In the crowd.'

I didn't reply. He looked over to the others, who were dancing. Esther, Madeleine and Zamya were really getting in to it, encouraged by Antônio. Even Rupert and Max were swaying from side to side. But I still didn't see what a packed bar full of sweaty, dancing students had to do with Esther's and my programming.

'Have you heard of the El Farol bar problem?' Alex asked.

I hadn't, but I did know that he was referring to the name of the bar we were standing in right at that moment, the one Chris had told Esther and me to visit. Chris had suggested we go here for a reason, Alex told me now, because the bar had an important lesson to teach us about stability and chaos.

I was still none the wiser, but, before I could ask what he was talking about, he pointed to the crowd. By now, the tempo of the music had increased, and everyone had their hands in the air.

'This is a small bar,' Alex said, speaking loudly now over the music, 'and there is space on the dancefloor for about fifty people. So, if forty or fewer people come to the bar on a Friday, then they have a great time.' Now, Alex went on, imagine that, because they had such a good time, the next week they each bring a friend, doubling the number of people at the bar. But now that there are eighty people at the bar there isn't space for everyone to dance. So, consider the following. The first fifty people find a spot on the dancefloor. Fine. Good for them. Each remaining person goes up and tries to take a space that is already occupied by someone else. An argument ensues between these thirty pairs of people trying to dance at the same place at the same time. As a result, the following week, each of these thirty pairs of people decide not to come back. Those who don't have their style cramped on the dancefloor invite a friend to come with them the next week, as before. The first question, Alex now posed, was how many guests come to the following week, then the week after that and the week after that? The second question was, in the long term, how many guests should the bar plan for each week?

The noise and the beer had softened my thinking somewhat, but it was straightforward to see that for twelve guests, for example, the next week there would be twenty-four. But for over fifty people, it was a bit trickier. So, I thought about what would happen if it was eighty people, as in Alex's example. Out of the fifty who occupy the dancefloor originally, thirty will end up in an argument with latecomers. As a result, only twenty (50 − 30) will have had a good night. The next week they would each bring a friend, so forty people would come to the bar.

I explained my reasoning to Alex.

'Exactly,' Alex replied. The method for finding the number for the following week is as follows: if the number of guests is less than fifty, we double it for the following week. For example, twelve doubled is twenty-four. If it is more than fifty, the number of 'excess' people who cause an argument is the number of bar visitors minus the fifty who got there first. So, in the example above, where there are eighty people, the excess is 80 − 50 = 30. We then take the excess away from the fifty people who first arrived on the dancefloor, i.e. $50 - 30 = 20$. This can be simplified by noticing that $50 - (80 - 50) = 100 - 80$. So, if we calculate for a hundred (the maximum possible number of guests) minus the number of guests and then double it, we get the number of guests for the following week. So, in the example, $100 - 80 = 20$, multiplied by two gives forty guests the following week.

I was now thinking about Alex's second question about how many people would visit the bar in the long term.

'I'd assume then that the number of people coming to the bar would stabilize at fifty,' I said.

It seemed obvious. When there are less than fifty guests, the number increases. When there are more than fifty, it decreases. It was something like supply and demand, the sort of problem Rupert would have scoffed at as trivial: the market will reach an equilibrium between bar patrons and available space on the dancefloor. I suggested this to Alex, and he just smiled.

I had fallen into his trap.

Think about the case, Alex said, in which forty-nine people come to the bar. They have a perfect night out and the following week they come back, each with a friend. Now there are ninety-eight people there, of which all but two end up fighting over space on the dancefloor. So, the next week there are only four customers! This is an extreme example, with a change from ninety-eight to four, but almost equally large hops eventually occur for pretty much any starting number of guests.

'You see,' he said, 'the problem is that you will never reach equilibrium. The numbers just keep going up and down.'

'Ah ha,' I said, thinking I now understood. 'It's similar to the predator–prey model, then? The number of guests is oscillating around fifty.'

Alex's smile broadened. 'No. Wrong again.' He went on, 'It is much more interesting than that.' This, he told me, is the starting point for chaos. However hard we tried, we would never be able to predict the number of people coming to the bar a few months from now, because if we made a small error in our measurement of the number of people coming to the bar, then this would lead to a large error in our predictions about the future.

'This is why economists like our friend over there don't understand anything,' Alex told me, pointing at Rupert. 'And it is why Parker's lectures so far are only a small part of the story,' he continued to explain. 'And it is also why calculations, like the ones you were doing with Madeleine for ants, only take us a small step towards the truth. Sometimes the reactions of life result in stable states or cycles, but other times they tell us something completely different. They tell us that life is uncontrollable. They tell us that we can't know the future. They tell us that we can't be responsible for our actions because we just don't know what the consequences are.' He told me that the question Chris had posed for us about chaos was only the beginning.

'But let's stop talking shop,' he said, finishing his drink. 'Just now, my unpredictable, chaotic logic is telling me that if we don't meet the locals, then we will never learn anything . . .'

With that he took me by the shoulder and dragged me in the direction of two women sitting at the other end of the bar. He greeted them with a big smile and introduced himself. 'Hello. I'm Alex. I come from Vienna, and this is my friend David from Manchester in England. Since we are from out of town, from Europe's two finest cities in fact, we were wondering . . . do you mind if we join you? . . . We are doing some research . . . do you come here often? . . . we were wondering if it is always as chaotic here as it is this evening?'

The chocolate cake of chaos

In Alex's bar problem, there is a form of self-regulation on the part of the El Farol guests. If the number of guests is less than fifty, then it increases (through positive feedback from bar patrons). If the number of guests is greater than fifty, then it decreases (it self-regulates). A similar idea of regulating or controlling a system underlies the way John treats his colleagues as basketballs: each push takes the basketball towards the middle. In John's case, these little pushes lead to stability. What Alex claims – but didn't provide me with a rigorous proof of in the bar – is that the result of the up-and-down regulations in bar attendance is not stability, but rather chaos. It is this claim which we are now going to explore.

Before we do this, let's look at another example of regulation by following a few months in the life of a cake-loving guy, Richard. Each year, since just before he turned thirty, Richard has put on a couple of kilos. He knows that he eats more sugary treats than he should. He feels like it is starting to affect his health. But he can't seem to get control of the situation.

It goes a bit like this. Richard decides to set a sensible limit for eating cake, pastries and desserts: once a week, which is four times a month. His resolution holds for a while but then, one week, someone brings a cake to work *and* there is a party at a friend's house where something irresistible is on offer. After he has broken the rules, he starts to give in to temptation more often, buying a *pain au chocolat* on the way to work or treating the kids (and himself) to a white-chocolate cheesecake on a Tuesday night. Pretty soon he is indulging every day, and soon after that it is twice a day. Six months later he realizes he is eating two or three sugary treats per day: a bun with his morning coffee, a cake or biscuits in the afternoon and a large dessert with his family in the evening.

He just gets carried away and can't stop . . . until . . . one day he wakes up feeling bloated, climbs on to the scales and realizes that he needs to change his ways. He decides he should only eat cake once a month, on special occasions, like a family birthday party. He makes a determined effort to go from nearly one hundred sugary treats per month to only one.

That feels much better, for a while . . . Until he thinks that it wouldn't be so bad to have a bit of cake every second week. Soon after, he tells himself, now he has learned a lesson, that maybe once a week is a good idea, like he tried before. That shouldn't be a problem, should it . . .?

We all have things which we know we should do in moderation – from eating chocolate cake to drinking whisky – but when we do it, we enjoy it, and can't stop doing it more. This is positive feedback: not positive in the sense that it is good for you, but positive in the sense that the more you do it, the more you want to do it. One indulgence a month doubles to twice a month; twice a month to four times a month; and one morning we wake up to the terrible truth that we are addicted! The opposite of positive feedback is regulatory feedback. This occurs when we suddenly decide to curb our consumption.

To help Richard (and ourselves), we need to look at the mathematical rule behind both Alex's description of El Farol and Richard's cake-eating.

Start by picking a number between zero and ninety-nine.

If the number is less than fifty, then double it. If the number is greater than fifty, then first take one hundred minus the number and then double it. So, for example, if you chose forty-five, then you get ninety. If you chose eighty, then the procedure gives $2 \times (100 - 80) = 40$. These are the same steps as Alex described in the bar.

Let's now look to see what happens when we choose two numbers which are close together and apply the rule to them. Starting with 13, we get the following sequence

$$13, 26, 52, 96, 8, 16, 32, 64$$

And starting from 14, we get

$$14, 28, 56, 88, 24, 48, 96, 8$$

The final number in the two sequences are very different. Starting with 13, we get 64. Starting with 14, we get 8. This is a form of sensitivity to initial conditions: the number you start with is all-important for predicting the dynamic to come. A difference of one at the start of the sequence becomes a difference of $64 - 8 = 56$ after only seven steps.

It is this small initial difference that is the hallmark of what mathematicians call chaos. Strictly speaking, we would not say that the sequences above are chaotic, since they both eventually repeat the same sequence $(8, 16, 32, 64, 72, 56, 88, 24, 48, 96, 8, 16 \ldots)$ – but the rule we have applied *is* chaotic. To see why, consider the plot in figure 11, which shows the sequence we get if we start with decimal values 14.1 (solid line) and 14.2 (dotted line). Initially, there is very little difference between the two sequences: the dotted line lying on top of the solid line isn't visible. But by step 9 there is a clear difference, with the solid line on 19.2 and the dotted line on 70.4. The two lines then spend a short time in sync again before, from step 14 onwards, they take their own independent paths.

The unpredictability of the outcome is not caused by any external random differences. Instead, it is the very small initial difference (0.1 in this case) that quickly grows, so that after twenty steps there is essentially no connection between the current value and the starting value. If I had chosen 14.01 and 14.02 as my starting points, then after about twenty steps they would have different values. If I had chosen 14.001 and 14.002, it would take about twenty-five steps for the differences to emerge.

Imagine now that the number sequence starting with thirteen describes Richard's cake consumption. In the story above, Richard doubles his eating of sweet things each month. He goes from thirteen to twenty-six, and twenty-six to fifty-two pieces a month. He does make a small adjustment when he finds that he is eating more

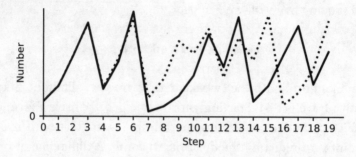

Figure 11: Two sequences of the number generated by the rule described in the text. The sequences, one starting at 14.1 (solid) and the other at 14.2 (dotted), move quickly apart.

than fifty pieces of cake per month and no longer doubles his consumption, going instead from fifty-two pieces of cake per month (roughly two pieces per day) to eat ninety-six pieces the month after. But then the truth dawns on him: he really needs to cut down. It is then that he goes back down to eight pieces per month (twice a week). Notice that the degree to which he cuts back depends on how extreme his consumption has become. At fifty-two he no longer doubled his consumption, but it wasn't until ninety-six per month that he really cut down.

Now think of the sequence starting with fourteen as an alternative reality: instead of eating thirteen pieces of cake in January, he eats fourteen pieces. Otherwise, he follows the exact same rule of increasing his consumption and then regulating it dramatically when he feels it has got out of control. In the first instance, he will end up eating sixty-four pieces of cake during the month of August (seven months later, or seven steps along in the number sequence). In the second instance, he will eat just eight pieces. Having one extra piece of chocolate cake at a party in January leads to a very different outcome in August.

Richard feels as if he is following a sequence of, for him, more or less logical steps. He knows he is allowing himself to drift into temptation, but when it gets too much he puts his foot down and goes on a diet. Nothing particularly strange about this. But to his friends, looking in from the outside, his behaviour is totally random. One summer he is taking them all out to cafés every afternoon; the next summer he refuses to touch even the smallest scoop of ice cream. He is caught in the chaos of cake consumption.

Alex's bar story, Richard's cake consumption and the number sequences have three key elements in common: positive feedback, regulatory (or negative) feedback and small perturbations. In the bar problem, the positive feedback is caused by people telling each other about the bar. The regulatory feedback arises when too many people visit the bar and fewer people go there the week after. The small perturbations are caused by little differences in the initial number of bar guests. For Richard's cake-eating, the positive feedback is the pleasure he gets from eating, which leads to more eating. The regulatory feedback is that he cuts back drastically when he is consuming too much. The small perturbation is the one piece of chocolate cake he did or didn't have in January. For the number sequences, the positive feedback is the doubling, the regulatory feedback is the downward adjustment made when the number exceeds one hundred, and the small perturbation is the difference of 1, 0.1 or even 0.001.

While we don't really expect Richard's cake consumption to follow the exact sequence of numbers obtained from the number rule (nor does Alex expect the number of people at a bar to follow the rules he describes), the way we consume and then sharply regulate ourselves has these same three elements: positive feedback, regulatory (or negative) feedback and small perturbations. Indeed, the chaos we saw in figure 11 does not depend on the exact rules. Mathematicians have found sensitivity to initial conditions for a wide range of rules in which small numbers multiply and large numbers become small. Whenever positive feedback is followed by a sharp regulation, chaos can occur.

The irony is that it is our attempts to regulate ourselves that are creating the chaos. We all fall into this trap: we decide not to use social media for a week; we stop drinking alcohol completely for a month; or we decide to get out running and immediately set off at our top speed around the park. All these extreme responses are a form of regulation or control. But they are also exactly the type of control that generates chaos.

By recognizing that regulation can create chaos, we can find an alternative method for getting things back on an even keel. When thinking about over-indulgence, we often tend to focus on the positive feedback: the fact that one slip will lead to a further slip, and so on. But this is the hardest thing to regulate, because it is fighting against our inherent tendency to lose control and want more.

A better solution is to aim to first stabilize and then slowly deflate behaviours we want to avoid. In Richard's case, under his current chaotic regime, his average monthly consumption is fifty pieces a month. If he could slowly cut his consumption down to, for example, one piece of cake a day (thirty per month), then he would be doing better than his current habit and it would be an easier level to hold. Allowing himself to have one indulgence per day is manageable. Each morning, Richard decides at which point in the day he will allow himself the treat: on his way to work, at afternoon tea, in the evening with the family, or even a sneaky snack after everyone has gone to bed. But he makes sure he is allowed only one treat, decided in advance. Gradual, carefully planned changes succeed, where drastic measures fail.

Recognizing the patterns in our own behaviour that can lead to chaos isn't easy. When we decide to make a big change in our lives – like clearing out a cupboard of old clothes, changing our exercise regime, giving up alcohol, making new friends, avoiding certain people, tidying our desks or setting up a new work routine – we do them because we want to regain control of our lives. These decisions appear to make sense on the basis of where we are in our lives today, but in six months we end up behaving in a way that is totally different than we expected.

The mistake

The best computer programmers, Margaret Hamilton believed, caught the mistakes before they happened.

Hamilton coded directly in binary, punching holes into the paper tape that was then fed into her LGP-30. There was nothing worse than the feeling she got when she made a mistake; she would sit on the computer-room floor, creating new holes in the paper and taping over the holes that had been incorrectly punched. When she started programming, there was always the awful uncertainty as to whether or not these modifications – this hacking, as she called it – would fix her original mistake. No, she thought, mistakes are to be avoided at all costs.

On one occasion, at three o'clock in the morning, Margaret was at a cocktail party with her friends when she suddenly realized that the computer would have finished its calculation and be sitting idle. She immediately went to Lorenz's office and started the weather calculations running again. When Lorenz arrived the next day and looked at where the calculation had got to, he realized that she must have been in early in the morning and asked her why. Hamilton told him that she had to do it. For her it was the obvious thing to do. She didn't want to waste even a few minutes of computing time.

Over time, the LGP-30 submitted to her control. She commanded it, and it responded to those commands. This was why it was all the more surprising when the discovery came: the insight that would make Edward Lorenz famous but would also shake Hamilton's beliefs to the core.

It started with a simulation Hamilton had made of a set of twelve atmospheric equations. She and Lorenz had, the day before, run the simulation and printed out a list of how the values of each of the variables changed over time. But when they reran the simulations the next day – inputting the same starting values, exactly as they stood on the printout – they got completely different results. The same input, but a very different output. They were shocked. How could this be?

At first, Hamilton worried that the problem was a programming error on her part. But she couldn't see where such a mistake might lie. Her rigorous approach, always working to prevent errors before they happened, meant that such an explanation was unlikely – but she knew that no eventuality could be ruled out. She went through the paper that had been fed into the machine again and again, trying to find out what had gone wrong.

And then they noticed it. The input to the code was specified to six decimal points, while the output on the printer was to three decimal points. The fourth decimal point, which they had failed to supply in the second simulation, had had a huge effect on the output. Despite the conditions being almost identical initially, the two forecasts were completely different. An input of 14.956 gave a very different output than 14.956181.

Edward Lorenz later referred to the effect as that of a seagull flapping its wings: the fourth decimal place was the sea bird, and, with the briefest movement of its wings, it had managed to change the weather (or, more accurately, the simulated weather . . .).

Margaret Hamilton was less interested in the importance of her simulation for weather prediction and more interested in the nature of the error itself. It had been a tiny difference in the input that had generated a massive difference in the output. She had, since her lessons with Professor Long, seen computations as logical and infallible. This time there had been no mistake with her code, but still an unexpected difference had arisen.

Lorenz praised her. He said that the reason he could see the problem, and discovered the seagull of chaos, in the fourth decimal was because he had so much faith in her programming skills. The input was the only possible reason for the error.

This was small comfort to Hamilton, who felt something akin to physical pain when things went wrong. She knew now that in the future she would have to be even more careful. She had big plans for her computer programming, and those would require even greater precision. Such mistakes, even if they were not her own, simply could not be repeated.

The butterfly effect

In 1961, Edward Lorenz switched his attention from the twelve equations which Hamilton had simulated on his lab's LGP-30 computer to a smaller set of just three equations which described processes of atmospheric convection. His aim was to capture the essence of weather systems in the simplest possible manner.

To get a feeling for the equations Lorenz was studying, imagine a tropical island. The ground of the island heats up as it absorbs radiation from the sun. As a result, the air on the surface rises and the colder air in the sky starts to sink. This produces a breeze on the island due to a convection cycle, where the warm air rises on one side of the island, to be replaced with cold air coming in as a breeze from the other side.

Lorenz captures this process in three mathematical variables: X is the intensity of the air convection or, equivalently, the strength of the breeze on the island; Y is the temperature difference between the east and west sides of the island; and Z is the distortion in the temperature profile between the island's surface and the sky above it (we say a temperature profile is distorted if the ground gets very warm while the areas furthest up in the sky get very cold). Note that the breeze will sometimes blow from east to west (in which case the variables X and Y take positive values) and other times from west to east (in which case X and Y take negative values). Lorenz proposed mutual feedback between X and Y: as the temperature difference between east and west increases, so too does the strength of the convection, and vice versa. This feedback leads to an increased temperature distortion between the land and the air, which in turn dampens (down-regulates) the convection, until at some point the wind switches direction and starts blowing from west to east instead. Over time, under the Lorenz model, the wind

146

will switch direction, blowing from east to west, then west to east across the island.

Like Richard's cake story and Alex's bar story, Lorenz's model is a simplification and my explanation of it in terms of the weather on a tropical island is also a caricature. But Lorenz equations for changes in X, Y and Z do capture many of the types of interactions we see in more comprehensive models used in weather prediction.

Margaret Hamilton didn't write the simulation of Lorenz's simplified model. By the time he had formulated it, she had moved on to another adventure (which we will soon follow her on). But before Hamilton left, she found her replacement: Ellen Fetter, who had also studied mathematics and had recently arrived in Boston. Fetter had much of the same attention for detail as Hamilton, and an ability to visualize how the output of the model could be communicated more clearly.

Fetter worked out how to get the LGP-30 to plot the output of Lorenz's model – the change in the variables over time – on paper. The plot she created was like the phase planes for the predator–prey and epidemic models which we looked at in the previous part of the book, where the lines represent trajectories of species or infections, but she plotted the change over time of three variables – X, Y and Z – rather than two. This is shown in Figure 12. To understand this plot, think of the three variables as travelling inside a cube. The height of the cube is the Z variable, while the X and Y variables are the two axes of the cube's base.

The trajectory continues round and round, but never repeats, in the sense that we never get exactly the same values of X, Y and Z twice. Instead, it forms a shape not entirely dissimilar to the wings of a butterfly as the weather on the island twists and turns in three dimensions. This is a very different type of dynamic than those that we saw earlier for the predator–prey model, for example. It is neither stable, settling down at a single point, nor is it periodic, repeating the same pattern over and over. First the wind blows from east to west, then it blows from west to east. Sometimes it blows warm, sometimes cold. And predicting the weather only a

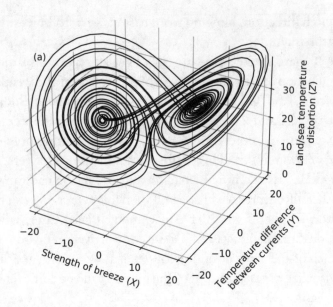

Figure 12: The butterfly of chaos. The line traces the change in the three variables over time.

few hours into the future becomes impossible. The trajectory is chaotic.

Although the concept of chaos has become synonymous with this butterfly, the fact that the diagram Fetter drew for Lorenz's 1963 article looks like a butterfly is just a happy coincidence. As we read in the previous chapter, Lorenz initially used the analogy of a seagull flapping its wings to describe the chaotic dynamic Fetter plotted. The more evocative image of the butterfly came later, when in 1972, Lorenz having failed to provide a title for a talk he would give, the organizers chose a question: *Does the flap of a butterfly's wings in Brazil set off a tornado in Texas?*

Evocative, yes, but this title also gives a slightly misleading view of chaos. It is not as if there is a single butterfly somewhere deep in the Amazon that through a flap of its wings causes a raging whirlwind in Texas. A more accurate way of describing the butterfly effect is to say that in order to accurately predict storms in the North Atlantic two months in advance, we need to know about air disturbances everywhere on the planet, and that includes knowledge of whether that Amazonian butterfly flaps its wings or not. In terms of figure 12, only a slight deviation in our trajectory takes us on another, completely different loop.

It is not the butterfly – or the one piece of chocolate or the one reveller at the bar – that makes the future uncertain. It is our inevitable failure to know about every possible butterfly, every piece of cake and every stranger at the bar that makes life unpredictable.

The night sky: part 1

We sat still and stared up at the stars. I didn't know how different the night sky here would be to the one I had grown used to in the UK. I certainly hadn't realized how bright the New Mexico stars might be. Back in Manchester, the sky was dimmed by a murky mixture of clouds and faded street lights. The rain and the cold made it unlikely that anyone would sit, as we were now, and look up at the night sky.

'You know they are organized to perfection,' Lily-Rose, who I was sitting beneath the stars with now, told me. 'But it isn't the type of perfection that you and your friend Alex are interested in: scientific perfection. It is a perfection that reflects ourselves back to ourselves, because, in our own way, each of us is perfect. I read that there are 10 billion galaxies out there and each of those galaxies contains 100 billion stars. Then I realized there will soon be 10 billion people on the planet and we each have 100 billion neurons in our brain. Each star, when it flickers, is a neuron firing, connecting to another neuron. Somewhere in someone's brain that connection is being made.'

It was when Lily-Rose understood this relationship – between brains and galaxies and neurons and stars – that she realized the power of astrology. Not the dumb astrology you read in the columns of the newspaper but the true astrology, the knowledge that has been handed down to us. When our ancestors looked out at the sky, she told me, they were able to read the firing of their neurons. Their own thoughts and the thoughts of others were written in the darkness and emerged from the flashing of the stars. Now that we had polluted the sky with our technological light and clouded our brains with scientific doubt, we could no longer see the patterns clearly. But, she told me, by coming up here she could understand the collective flashing of our minds.

Alex and I had been brought to this point, in the foothills of the mountains, a short distance from Santa Fe, by Lily-Rose and her friend, Maria – who we had met earlier at El Farol. Lily-Rose had driven the car. I sat next to her, in silence, trying to ignore the giggles and whispering coming from the back seat, where Alex and Maria sat close together. When we arrived at our destination, Lily-Rose and I both got out, leaving the newly formed couple in the car, and walked a short distance to the vantage point at which we were now sitting.

It was only then that she started talking and told me about the stars.

I liked the imagery of her idea but told her that it sounded quite unlikely from a scientific point of view.

She responded that she had been to a public lecture at the Santa Fe Institute last year. The lecturer had presented what he called, if she remembered rightly, cellular automata. He had talked about how everything was connected, how a small change in one place in his automata could lead to a large change elsewhere. He had said that the universe was the same, a flicker here was a change in a galaxy on the other side of the universe. As she understood it, cellular automata could be used to model our brains, the stars and everything else as an array of interconnected flashing light bulbs. One bulb would flash and set off another. The firing of our neurons and the flickering of the sky had unpredictable effects.

That is what we are seeing now, she said, the flickering of our minds in the sky. Everyone has their own star, which represents what they are thinking about. We see the chaos of our thoughts. This was why astrology is true, she repeated.

'When we say that it is written in the stars,' she told me, 'people like you and Alex – scientists – take it literally. But that's not the point. The point is that it requires a special feeling to read the stars, a special feeling to understand our minds and a special feeling to read the future.'

I wanted to tell her that this probably wasn't really what the lecturer had meant. I knew that while Chris (it sounded like it might have even been Chris who gave the talk she had attended) might say that both the universe and our minds could be modelled as zeros and ones, like the flashing of light bulbs, he didn't literally believe in a causal connection between brain cells and stars.

But while I was thinking how to best explain this to Lily-Rose, she continued, 'I know what I said shouldn't be taken literally. But I have had a lot of chaos in my life – people near me who are out of control. And thinking in this way helps.' She explained that this was why she had liked the talk by the scientist, because the speaker had said the same thing: that we might think that one thing causes another, and on one level it does, but when we stop focusing on the everyday and look instead at the flashing light bulbs, the patterns are random.

'Yet most of us never come up to the mountains to look into our own minds,' she said.

Coming up here reminded her that she didn't need to be in control. That other people were like swirling galaxies, and they would continue on their own trajectory, irrespective of what she did or didn't do. It made it all just that little bit easier.

The night sky: part 2

For Margaret Hamilton the night sky meant only one thing: the need for total control. It couldn't be any other way. It was a vacuum, an empty space, governed only by the laws of gravity. The success of the mission was entirely determined by the rocket trajectories she and her new colleagues at NASA decided upon. Their job was to make sure the mission succeeded, that *Apollo 11* made it through that vast emptiness and successfully landed astronauts on the moon. And that it came back again. Even the smallest of errors was out of the question.

She knew, from the chaos she had seen in weather predictions, that even a single small mistake, like in the fourth decimal point, could result in failure. This lesson had been reinforced in her work after she had left Lorenz's lab and the LGP-30 behind and found employment at American Homeland Security, writing software to detect unfriendly aircraft. She named her new system 'the seashore', because of the beautiful, regular sound the large mainframe computer would make when it was running her software correctly. If she had got something wrong, the sound would change from one of gentle waves washing against the sand to the gale of a fierce, unpredictable storm. The worst possible outcome was a computer crash, which would cause sirens and foghorns to blast out, telling the world she had made a mistake.

Hamilton learned from her earlier mistakes. When they did occur, she would look for new ways to classify errors; document everything. She would take Polaroid pictures of her colleagues posing next to their bugged code. It was better when a large audience witnessed her mistakes, she thought.

And this was why she knew, as soon as she heard that NASA

needed someone to write the first 'man-rated software' – code that could be used to safely send an astronaut to the moon – that she was the woman for the job. She had heard Kennedy's words – 'We choose to go to the Moon in this decade and do the other things, not because they are easy, but because they are hard' – and she understood just how hard it would be. Just how little room there would be for error.

In 1963, Hamilton applied for computer-programmer openings available within two different NASA teams, and both of them offered her a position within hours of the interview. Ironically, she would leave it all to chance and let a coin toss decide which team she joined. It didn't matter in the end, because it was soon obvious to everyone at the space agency that Hamilton should be involved in all parts of the mission to the moon. There had, up until she arrived, been a culture within NASA (and in computer programming as a whole) to place value on intricate and complex code. What Hamilton gently but firmly pointed out was that the more equations that were involved, the bigger the probability that an error would slip in. She pushed for simplicity, repeatability and understandability. She coined the term 'software engineering', because the work they were doing was, she believed, just as important (or even more so) as that of the engineers who built the rockets or assembled the lunar modules. To be as far away from failure as possible, the software they engineered should, she argued, be sleek and streamlined. Just like the spacecraft itself.

The key to Hamilton's approach went back to her lessons with Professor Long in college. Long had shown Margaret that allowing each point to follow logically from the last was a much more certain path than memorization. The same idea could be applied to computer software. The computer on board *Apollo 11* was responsible for many vital functions, including estimating the position and velocity of the spaceship, providing assistance with steering commands, controlling the temperature of parts of the ship and helping the astronauts measure angles between planetary bodies. Each task had a different priority in each different situation the mission faced. For example, the steering function initially had priority over estimating position, but if the ship had moved long distances without a measurement being made, then its position became uncertain and re-estimating position increased in priority. Hamilton's task was to develop a system that allowed the onboard computer, which could only compute one thing at a time, to deal with the highest-priority task first.

Instead of writing a list of 'if . . ., then . . .' procedures to deal with each potential eventuality, Hamilton first documented the function (the role) of each part of the spacecraft controlled by the computer and the priority relationships between those functions. She and her colleagues then built a software system which, given a list of functions and priorities, would automatically produce the correct response. Once they were 100 per cent sure this overall system was failsafe, they knew that if a particular response was incorrect in practice, the error lay in the way the function and priorities were described. These errors were much easier to debug than those hidden within 'if . . ., then . . .' statements, because they related to the functioning of the spacecraft and the priorities of the mission. Furthermore, if the functions of the spacecraft or the priorities of the mission changed, which they often did during the eight years of the project, updates could be made without the risk of introducing new errors.

Hamilton's approach to software took inspiration from the way engineers write down equations that describe how instabilities can arise within a system. For example, equations that describe how a suspension bridge – like the Golden Gate Bridge in San Francisco or the Millennium Bridge in London – can start to oscillate. By understanding where instabilities lie, the engineers can (in most cases) control a system and keep it stable. For Hamilton and her team, the uncertainties could be found in the high power thrust of the spaceship's burners, the small error in the spaceship's positional measurements, or a miscalculation by an astronaut or by mission control . . . These are the butterflies in the space mission that needed to be identified.

Before they created chaos. The eight years of software engineering Hamilton and her team undertook was about planning for each and every one of the eventualities. They had to control the chaos long before it ever had the chance to arise.

Margaret Hamilton was in the control room throughout the *Apollo 11* mission, checking the monitor and reading the printouts. She watched carefully as the astronauts' descent to the lunar surface began. It was a critical moment: the last obstacle before humans walked on the moon for the first time.

It was exactly then that an alarm rang and lights started flashing at mission control. On the astronauts' computer a warning appeared. One they had never seen during training: an emergency code which indicated that the computer was overloading. Armstrong's voice to mission control sounded concerned: 'It's a 1202 . . . What is that?'

All the engineers in the room turned to Hamilton. It was *her* software that had raised the alarm: the 1202 routine had interrupted the astronauts' mission display to warn them of an emergency. What was wrong with *her* software, they wanted to know.

On board, Armstrong and Aldrin saw a message on their mission display. It told them they needed to manually return the rendezvous radar to the correct position before landing. They flicked the switch and moved it back to where it should be. The display now asked the astronauts whether they were ready to land. They made the decision to go and started the final descent. The alarms and warning lights switched off.

'What happened there?' asked one of Hamilton's colleagues nervously.

Hamilton looked calmly down and checked the printout. It wasn't a problem with her software. It was the spaceship's hardware that had failed! Her computer code had both compensated for the error in the hardware and alerted Armstrong and Aldrin to the problem.

While the others cheered the first humans to land on the moon, she thought to herself, 'And also the first software to land there too.'

She smiled to herself as she thought of that small, fault-free box of computer routines resting on the lunar surface.

Many decades later, when Barack Obama awarded the Medal of Freedom to Hamilton at the White House, he introduced her by saying, 'Our astronauts did not have much time, but thankfully they had . . . Margaret Hamilton, a young MIT scientist and a working mum in the sixties.'

He reminded his audience that Hamilton had been engineering software before the term was even introduced. 'There were no textbooks to follow,' Obama said, 'so there was no choice but to be pioneers.'

Obama was right, Hamilton thought as he spoke. She didn't have a textbook. But she had been taught to approach her work in a way that encouraged rigour, that used logic and reasoning to eliminate mistakes before they happened. She had witnessed the first chaos in computer simulations, emphasizing for her that the software engineering on the Apollo mission would have to be perfect. She had taken her own deepest fears of getting it wrong, of erring when certainty is needed, of letting life slip out of control, and she had used them to transcend the heavens.

Her mission was complete.

The perfect wedding

Margaret Hamilton at NASA and Lily-Rose in Santa Fe provide two very different approaches to chaos and randomness. Hamilton's solution to avoiding chaos was to take preparation to its far edge: eight years of planning were needed to secure safety during the few minutes of the descent to the moon's surface. Lily-Rose's approach recognizes the impossibility of always controlling our lives. It asks us to accept chaos.

In our own lives, the challenge is to know which approach to take and when.

To find this balance, let's get to know Nia, who works as a wedding planner. In fact, she is one of the UK's most sought-after wedding planners. She has recently featured in a reality-TV show, *My Big London Wedding*, and business is booming as a result.

Nia once worked as an investment banker, after studying engineering at university. Three years ago, she quit her job to put her technical skills to the ultimate test: to sort out the logistics for the most important day of many people's lives.

The show follows Nia's step-by-step journey with the happy couple (and often the parents, who are picking up the bill), from explaining the details of wedding flowers to planning the menu. On the big day she is at the venue at 6 a.m. and stays until the dancing is underway. She communicates with her assistants using walkie-talkies, coordinating everything from elaborate light shows to hair and make-up. She has never had a serious problem: the cake is always how the bride and groom want it, the limousines are always on time. She loves that moment, the culmination of a perfect day, when people are invited into the dining room, and they stop for a time at the door to take in the sight, to coo and take pictures. It is a moment of true perfection.

Nia feels differently, though, when she comes home. Because of the hectic nature of her job, her husband, Antony, takes primary responsibility for childcare and the household. He is happy with his role as an active father, and just as she loves her job, he loves his. But the thing is, he isn't very good at it! Well, he is good with the kids and always finds new ways to keep them entertained. They are always starting (and not always finishing) creative projects: one day they'll be painting, the next taking part in a sports tournament, or perhaps a board-game marathon.

Not only that, Antony often has their friends round, who can be worse than the kids. When she came home last week, they had started a project to study data science and try to understand their happiness using statistics. Arriving through the front door, she finds Antony, Aisha, Charlie and Becky huddled over their laptops in the kitchen, while the kids are jumping up and down like crazy in the living room.

When Nia comes home, she can't relax. Antony does tidy up . . . eventually . . . late at night, once the kids have gone to bed and their friends have gone home. But she doesn't want to have to come home to chaos. What should she do?

The answer to this question can be found in the difference between Margaret Hamilton and Lily-Rose. Like Hamilton, our wedding planner controls that one day to perfection: she carries out a moon landing every weekend. In organizing a wedding there is no room for error. We need Hamilton's decimal-point precision and perfection engineering: every eventuality should be planned for; even the smallest potential bug should be eliminated in advance. This is how Nia works too. Everything must be perfect on the big day and that requires a plan that can deal with every possible hiccup.

What Nia cannot control, though, is what happens the day after or the day after that, how the couple's marriage develops. This is the nature of chaos. This was the lesson Lorenz learned in 1961, when he and Hamilton looked at the outcome of the simulation of the atmospheric equations, where the tiniest of errors made results unpredictable. He realized that we can control or predict the future only for a very short period of time – we can carefully regulate the landing of a spacecraft or even give a (relatively) accurate forecast of whether it will rain in the afternoon – but the further future is not something we can anticipate. Chaos is inevitable.

Our wedding planner is remarkable because of her ability to control the future in the short-term, but she cannot take control in the longer term. When Nia looks at her husband, she needs to understand a different way of thinking, a way which accepts that nothing can be perfect in the long term. That said, Antony also needs to understand Nia's way of thinking: the view of stability and perfection. It is not that one is wrong and the other is right. They both have their place.

In Chinese philosophy this dichotomy is called yin and yang. The yin is chaotic; it is passive and allows itself to flow into the unknown. The yin is Antony. He allows himself to be dragged around by whims and desires. The yang is order; it is active and aims to control the future. The yang is Nia when she is at work. Every second is under her control.

Nia and her husband need to balance her yang with his yin, her short-term order with his long-term chaos. In practical terms, this means the couple should talk about which aspects of their lives they want to control tightly and which they want to allow to flow freely. For example, Nia says that she can accept some chaos from the kids. Both Nia and her husband agree that children need a daily structure, regular mealtimes and night-time routines but also need to have the possibility to express themselves freely without constraint. In a similar way that Nia enjoys seeing her wedding guests letting loose on the dancefloor, she can accept that her husband allows the kids to let loose, to learn about chaos in a safe environment.

Disorder in the kitchen, though, is another thing. It is here that Antony's chaotic yin has gone too far. Nia's orderly yang has nowhere to breathe and recover. She needs at least one place in the house where she can relax, where there are no toys or half-finished art projects left lying around or data-science enthusiasts tapping away at their computers. A place where the two of them can prepare an evening meal or enjoy a glass of wine together (after the kids have gone to bed). They decide that order in the kitchen is a priority and Antony promises to keep the adult spaces tidy. He also agrees to meet with his friends in the evening outside of the house, allowing Nia to spend a bit of time with the kids. They agree that if he can't cope with both the kids and the housework (it can be tough), then they will pay for cleaning help, get a babysitter or order (healthy) takeaway food more often (just as Antony and his friends' statistical analysis in Part I recommended).

While the exact details of finding that balance of yin and yang depends very much on the details of a relationship, chaos theory proves that we cannot have one without the other. Even an incorrect fourth decimal place can completely change a carefully calculated result in the long term. Order and chaos are intimately intertwined, like the lives of married couples. The key is to recognize that trying to control the long term leads to over-regulation and even more chaos, but neglecting to control the short term leads to insecurity and even less order. Getting that balance right isn't easy, but recognizing that neither order nor chaos can live without the other is a good start.

This balancing act leaves us with a new question. While we have learned about the yang side of the balance through studying stability and periodic systems and following Margaret Hamilton as she eliminated unpredictability, we are yet to look at the yin side of the balance.

If we are going to allow ourselves to let go, like Lily-Rose does, and give in to the chaos, what do we expect to find there?

To find out, we need to go back to Santa Fe.

Cellular chaos

I slept until after lunchtime on Sunday, but when I did wake up I was keener than ever to do the exercise that Chris had set us: to find chaos in an elementary cellular automata model. After learning about chaos from Lily-Rose and Alex, it felt like I had new inspiration. I decided to try to find Esther, so we could do the exercise together. But when I got to the common room the only two people there were Antônio and Madeleine. They were, despite obvious hangovers, arguing about the details of ant and wasp evolution. Madeleine told me that I had missed Esther. She, Rupert and a few others had gone to the open-air Santa Fe Opera.

So I went to the computer lab by myself, sat down and started programming. It was difficult to think in a way that was random. I recalled a magic trick I learned as a kid which involved asking a friend to 'pick a number: one to four'. Most of us intuitively pick 'three'. I had the same problem now. All the ideas that came to my head were regular and periodic. I just kept picking threes.

While I was trying out some different rules, I remembered that Esther had explained to me that all elementary cellular automata could be written as a set of rules that looked like this:

111	110	101	100	011	010	001	000
0	0	0	1	0	1	1	0

Remember, a cellular automaton is a set of rules for making changes to a string of 1s and 0s, a binary string of bits. For example, consider a starting string of

00000001000000

To see how we apply the rules, start by looking at the 1 in the middle of the string. It has two 0s as neighbours, and so the pattern they form is 010. Looking up 010 in the rules above, we see that it gives a value 1, so in the new string the 1 will remain a 1. Now, if we look at the bit to the left of the central 1, then the neighbourhood pattern is 001. Looking up the rule for a neighbourhood 001 tells us that the 0 should change to a 1. Similarly, the neighbourhood to the right, which is 100, also changes to a 1. Applying the above rules gives us

00000011100000

The central 1 has become three 1s (note that the parts of the string with 000 remain 0, following the rule furthest to the right above). Applying the rules one more time, we get

00000100010000

This is because, if all three bits are 1 (i.e. 111) or two adjacent cells are 1 (i.e. 110 or 011) then, according to the rules, the bits become 0.

I implemented these rules in code and ran them on the computer, starting with a single black cell in an array of white cells. The black cells in this case represent 1s; the white cells are 0s. I watched as the cells filled out the screen one row at a time.

I recognized the shape it produced from maths classes (figure 13). It was a fractal, a self-similar pattern. The overall triangle created by the cellular automata contained three smaller triangles, each of which in turn contained three smaller triangles, and so on. When I learned about this particular fractal, known as the Sierpiński triangle, in a mathematics course, we were shown how to construct it by starting with a black triangle, then colouring the middle of the triangle white, after which we would colour the middle of the remaining three black triangles white, and so on. But here, the same shape had appeared from a completely different method. The elementary cellular automata had built the Sierpiński triangle using a simple set of binary rules.

Transition
Rules

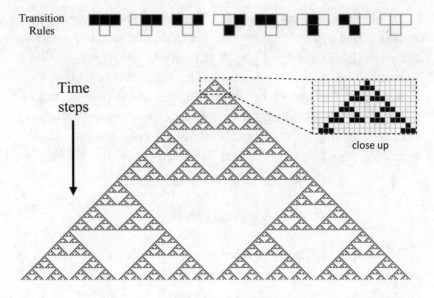

Time
steps

close up

Figure 13: A cellular automaton which produces a fractal-like pattern. The top row shows the transition rules for how the three neighbours on the row above determine the row below. The evolution of the cellular automaton as time changes from top to bottom.

I was inspired by the way symmetrical beauty could arise from such a simple rule, and continued to play around with different rule sets. It was then that I found it. Just one rule change (so that 011 went to 1 and not 0) gave a rule set

111	110	101	100	011	010	001	000
0	1	0	1	1	1	1	0

Could a small change to the rules, one bit out of place, lead to a large change in the output?

Yes, it could. Running this simulation, again starting with a single black cell, gave a completely different pattern. On the left-hand side of the plot (figure 14) there appeared to be some semblance of regularity. Small, repeating patterns. Regularly spaced triangles. These triangles varied in size, larger on the left than the right. The right-hand side of

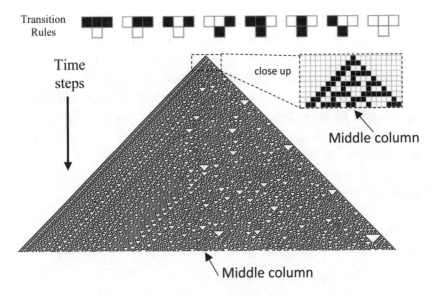

Transition Rules

Time steps

close up

Middle column

Middle column

Figure 14: A cellular automaton which produces a random pattern. The top row shows the transition rules for how the three neighbours on the row above determine the row below. The evolution of the cellular automaton as time changes from top to bottom.

the plot was very different: regularity is lost and randomness dominates. Big white triangles, small white triangles and lines intermingle without any clear order. The randomness appeared most extreme in the middle of the shape. There were just as many white as black cells and there seemed to be no way of predicting what would come next.

This was it, I thought, a random pattern, generated by simple, deterministic rules.

Chris was enthusiastic when I showed it to him at the lab on the Monday afternoon.

'Humungous,' he said. 'You got it!'

We both looked at the changing patterns on the screen as the simulation continued to run.

'How do I know it really is random?' I asked. 'It seems most random in the middle, but I don't know how to measure it.'

Chris told me that this was the right question to ask, so I waited, expecting an answer to come. But it didn't. Instead, after we had watched the pattern scroll by one more time, Chris said just a single word to me: 'Entropy.'

That word again. The one Max had, on our first evening in the sports bar, used to describe the USA's addiction to information. Entropy had something to do with communication. But what? Chris wouldn't say. He wondered off to talk to another student.

I was going to have to work out entropy myself . . . or, better still, maybe I could find Esther and get her to explain it to me . . .

A message from B to C

Hamilton provided us with a yang, that of controlled engineering stability, but we still need a mathematical yin, the secrets of entropy. To find this, we need to go back in time again, to 1948, and meet a young man trying to overcome his shyness.

The question as to whether Betty would go on a date with Claude had consumed most of his thinking for the past few weeks. Her answer contained only one single bit of information: a zero-no or a one-yes. It was ridiculous then, he thought, that so much of his own computational machinery had been fruitlessly dedicated to predicting the well-defined binary configuration hidden within Betty's mind.

Claude started to think about how his own theoretical findings might be applied to the problem. His work, at Bell Telephone Laboratories in New Jersey, had led him to propose something that he called communication theory. The question that motivated his theory was about the best way of converting letters of the alphabet into binary 1s and 0s so they could be sent efficiently between two points along a communication channel. How could telecom engineers best encode a message between two points? Between point B (Betty) to point C (Claude), for example.

The theory told him that her binary 'yes' or 'no' was an accurate and concise answer. An efficient communication. His own internal speculation about what her answer might be, on the other hand, was ridiculously inefficient.

The best action, according to his theory, was therefore clear. The inefficiency in his thought needed to be replaced with her concisely encoded answer.

So, Claude Shannon finally asked Betty Moore if she would go out for dinner with him.

And her answer was a binary 1; an affirmative 'yes'.

This had not made things easier, he thought now, as they sat together in the restaurant. He realized that he had forgotten to think through what the single 'yes' bit ultimately entailed in polite society, namely, further communication in the form of two-way pleasantries. The irony was not lost on Claude. Despite having published what his peers at Bell Labs considered to be the most important paper ever written on the theory of communication, it was the very skill of communicating with others which he was sadly lacking.

He knew he should say something about the decor, or the food, or maybe even her appearance, but this was the problem. He simply didn't know how to talk about such things. Small talk was a waste of communication bandwidth.

But then he realized something: she had not said a single thing to him either. And she didn't, as far as he could tell from the thoughtful expression on her face, seem particularly concerned about the lack of exchanged words. She just sat there, observing him.

'Why aren't you saying anything?' Claude asked at last, once he had solidly determined that he had no other way of working out her silence.

'I have read your paper,' Betty said, 'on entropy, information and communication. I have a few questions and I wanted to formulate them clearly before I pose them. You left very little unstated in the paper, and I don't want to unnecessarily repeat arguments, but it might help if I summarized my understanding first . . .'

This was not the reply Shannon had expected.

Betty outlined, based on what she had read in Shannon's paper, how the first step in understanding communication was to realize that everything could be encoded in the 1s and 0s of binary. For example, if we want to encode the first four letters of the alphabet, we might say that A is 00, B is 01, C is 10 and D is 11. Remember, a bit is a single 1 or 0, just like a digit in a number between zero and nine. We could do the same with decimal numbers, writing 0 for A, 1 for B, 2 for C, and so on, up to 25 for Z (it is 25, not 26, because we used

0 for A). But we choose binary because it is the way data is sent along a cable in the form of two different voltages, with one voltage representing 1s and the other 0s.

A binary string consisting of two bits allows us to encode four letters in binary. To encode eight letters we need three bits (A as 000, B as 001, C as 010, D as 011, E as 100, F as 101, G as 110, H as 111). For sixteen letters we would need four bits, and so on. In general, each extra bit doubles the number of bits that we can encode. (Today, the ASCII code uses eight bits, known as a byte, to encode $2^8 = 256$ different letters and characters.)

'Now imagine,' said Betty, 'that we send a message consisting of a sequence of just the first four letters – A, B, C and D – with each letter in the sequence chosen at random.' She leaned forward and wrote on her paper napkin

BACDABACDDADBCCB

Each letter in the sequence occurs equally often: four As, four Bs, four Cs and four Ds. Then she wrote the sequence in binary – replacing A with 00, B with 01, C with 10 and D with 11. In this way, the string of letters was converted into a binary string of 1s and 0s, as follows:

01001011000100101111001101101001
B A C D A B A C D D A D B C C B

The sequence of sixteen letters required $16 \times 2 = 32$ bits to represent it. 'That's right, isn't it? You had a similar example in your article, didn't you?' asked Betty, looking up at Claude.

He just nodded in reply, waiting for her to continue.

'Now let's look at another example similar to the one you used,' she said. 'Imagine that the letter A occurred in the message most frequently, half of the time, B occurs a quarter of the time and C and D only occur an eighth of the time.' She wrote down a new sequence on the napkin, as an example

ACAABBABDABAACDA

It contains eight As, four Bs, two Cs and two Ds. One possibility would be to encode this in the same way as above, replacing A with 00, B with 01, C with 10 and D with 11. This gives

$$0010000001010001110001000010100$$
A C A A B B A B D A B A A C D A

And again uses $16 \times 2 = 32$ bits in total.

'But you like things to be concise and efficient,' she said, 'so you wanted to make this string shorter.'

This was precisely the problem that Claude had wanted to tackle. The above encoding was inefficient. In particular, the As in the sequence fill the string with an unnecessary amount of 0s: twenty-two out of the thirty-two bits in the string above are 0 and only ten of them are 1. Was there a way of sending fewer redundant zeros?

The key to efficiency is to find shorter codes for the letters which occur most frequently, Betty said. For example, if we write, A as 0, B as 10, C as 110 and D as 111 now, then the binary encoding is

$$011000101001011101000110110$$
A C AAB BAB D AB AA C DA

This code contains only twenty-eight bits (fourteen 0s and fourteen 1s), but still carries all the information from the original sequence of letters. Provided we know how the rules for the binary code operate, we can always reconstruct the sequence of letters.

'And this was how you came to entropy, isn't it?' Betty said.

The entropy, she explained, is the average number of bits required to send a single letter. For the first string, we needed thirty-two bits in total, or two bits per letter sent (there are sixteen letters in the string, so $32/2 = 16$). For the second string, we only needed twenty-eight bits, which is $28/16 = 7/4$ bits per letter sent. The entropy of the first message was greater than the entropy of the second message (because $2 > 7/4$).

A message where all letters occur equally often contains more information than a message with lots of repetitions of the same letter because we can't find a shorter encoding for the former. The first string Betty wrote down contains more information than the second string. And so, entropy is a measure of the amount of information in a string.

She sat back in her chair and looked at Claude.

'I hope you don't mind me repeating back to you a message that you have already broadcast to everyone at Bell Labs?' she said with a smile.

He didn't mind at all. She had put it even more concisely than he himself ever could.

It was the most beautiful communication he had ever received.

Information equals randomness

On the Tuesday afternoon, I went to the library at the Santa Fe Institute, to see if I could learn more about entropy. I found a copy of Claude Shannon's book *A Mathematical Theory of Communication*. Written in 1948, it had come far before its time. The central idea was that all data sources – the texts we write, the pictures from digital cameras, digitized music files on a CD and films on DVDs, even recordings of the things we say – can be seen in the same way: as a stream of 1s and 0s. The information contained in a data source, Shannon's book explained, was equal to the entropy, which was the number of bits required to code it in binary.

I understood the idea of information as coding letters in binary; what I didn't understand was what this had to do with randomness.

It was getting late, and there were only a few other students in the library. Esther was one of them, sitting on the other side of the room, focused on her research work. We hadn't spoken since the lab the Friday before. The last time I'd seen her was in the El Farol bar on the Saturday evening, when I had left with Alex, Maria and Lily-Rose.

After a while everyone else left and it was just the two of us in the library. So I went over and, rather tentatively, sat down at the desk she was working at.

She didn't seem particularly pleased to see me.

'I'm surprised you're here,' she said, looking me up and down. 'I thought you disappeared off in the evenings with the locals. Like on Saturday. It is amazing that you could do the homework Chris set, with all the distractions you seem to have found in Santa Fe.'

She was making light-hearted fun of me, but I somehow couldn't stop myself from blurting out the whole story of what had happened that Saturday evening, what Lily-Rose had said about chaos and the stars and our brains and the flickering, and so on.

'Oh dear. You come to Santa Fe to study and learn, and you end up smoking dope with a hippy girl and listening to her half-baked theories about the meaning of life,' she said.

'I suppose I am a bit confused,' I admitted. 'I want to know what randomness is. I can see when things are random and out of control, but how can I measure it? Is it even possible to measure it?'

'Real-life randomness or just randomness in your cellular automata simulation?' Esther smiled.

Without waiting for an answer, she continued, 'While your new friend Lily-Rose might think that chaos means that we have to give in to mysticism, that is not completely true.

'What you have missed, David,' she said, swivelling round on her stool to face me, her knees almost touching mine, 'is that randomness is information.'

'That is what Chris said!' I exclaimed. 'Or, rather, he told me that I should look up entropy. That's why I came to the library, to read Shannon's work.'

'And what have you found out?' Esther asked.

I explained that I got the idea that entropy is the average number of bits needed to send a string of text. What I still didn't understand, I said, was what it had to do with measuring randomness in my cellular automata. Nor could I see how it related to the type of randomness in our lives that Lily-Rose had referred to. What is the relationship between randomness and information?

Esther turned her seat back towards the desk and took out a piece of paper. She wrote down the two strings of letters (the same as those written by Betty on a napkin).

BACDABACDDADBCCB

and

ACAABBABDABAACDA

'Which of them is more predictable?' she asked.

I thought for a while, and realized that the second string was more predictable than the first, because it contained more As. If I was asked to guess the next letter in the sequence, then by guessing A I would be correct 50 per cent of the time for the second string and only 25 per cent of the time for the first string.

'Exactly,' said Esther. She reminded me, just as Betty had shown in her example, that the binary encoding of the first string required thirty-two bits, while the binary encoding of the second string required twenty-eight bits. In general, the more unpredictable the string is, the longer the binary string needed to represent it. It is in this sense that randomness is information: random strings require longer binary encodings, because they contain more information.

To see why this is, think about the average length of the encoding of each letter in the string. For the first string, each letter required two bits to encode and occurred one quarter of the time. So, the average length of the encoding for each letter is

$$\frac{1}{4}\times2+\frac{1}{4}\times2+\frac{1}{4}\times2+\frac{1}{4}\times2=\frac{8}{4}=2$$

bits. While in the second example, encoding an A takes only one bit and As occur half of the time, encoding a B takes two bits and Bs occur a quarter of the time, and encoding a C or a D takes three bits and they each occur an eighth of the time. Thus, the average encoding for each letter is

$$\frac{1}{2}\times1+\frac{1}{4}\times2+\frac{1}{8}\times3+\frac{1}{8}\times3=\frac{7}{4}$$

bits.

Esther went on to explain that we could think of the first string as if it had been created by throwing a four-sided die (a die built as a pyramid of four triangular faces). A die is completely random when every side is equally likely, i.e. has probability 1 in 4 as in the first string. A weighted die that is more likely to land on one specific side

174

is less random. It is more predictable. The second string is therefore like a die that is more likely to land on one specific side. More predictable strings contain less information.

The relationship between entropy and information is true, not just in this example, but in general, Esther explained. As an extreme, consider a die which always lands on the same side. The outcome of the throw provides us with no new information whatsoever. We know the result in advance. Similarly, the string

$$AAAAAAAAAAAAAAAA$$

contains no new information. It is entirely predictable. It has an entropy of zero.

'It is by using entropy that I can answer both of your questions: the one about your cellular automata simulation and the one about how to find the meaning of life in a chaotic and random world.' She smiled.

Esther asked me to write down the middle column of the printout I had made of the cellular automata (the process of doing this is illustrated in figure 15). I copied down the sequence of black and white cells as 1s and 0s as follows

$$010000110...101$$

She told me that even though the process which had generated this middle column was deterministic (it came from my cellular automata), there was no way for me to guess, just by looking at the bits which came previously, whether the next bit in this sequence should be a 0 or a 1. This meant that the entropy of this binary string was maximal, the middle column was completely unpredictable.

I listened, but by now I had lost interest in the cellular automata. I was waiting for the answer to the second, more important, question: how does entropy give us insight into the real world?

When Esther finished talking about cellular automata, she looked directly at me. There was a long pause. A silent anticipation.

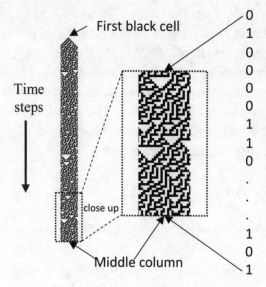

First black cell

Time
steps

close up

Middle column

0
1
0
0
0
0
1
1
0
·
·
·
1
0
1

The sequence of 1's and 0's is constructed by successively zooming in on the middle column of the cellular automate in figure 14 and writing a 1 whenever there is a black cell and a zero whenever there is a white cell.

Figure 15: To find randomness in the cellular automaton in figure 14, we zoom in on the middle column, the one that contains the first black cell. This contains a random sequence of 1s (black) and 0s (white).

'The way to deal with chaos and randomness in our lives is . . .' she said, moving her chair ever so slightly closer to mine, '. . . the game of twenty questions . . .'

With those words, before I could ask her anything more, or even properly comprehend what she meant, Esther pushed her chair back and stood up, saying '. . . and your permitted questions, David, have just run out.'

With that, she walked out of the library.

Twenty questions

Let's see if we can work out what Esther meant when she alluded to the game of twenty questions, the one where I think of an object and you have twenty 'yes' or 'no' questions to work out what that something might be.

To get warmed up, let's think about the number-based version of twenty questions. I think of a number between one and twenty and you have to guess it using a series of 'yes' or 'no' questions. This time I state that the number must be a whole number, and one and twenty are included. So, there are in total twenty different numbers to choose between. What is the best way to find out my number as fast as possible?

You can, if you want, just guess at a number – 'Is it fifteen?' But with a very high probability (19 out of 20, to be exact), I will have chosen something different and, when I say, 'No, that's not my number,' you will still be left with nineteen alternatives. In the worst-case scenario, you will have to use all twenty questions to get to the answer.

A better strategy is to use 'more or less' questions. For example, 'Is it more than fifteen?' you might ask. If I answer yes to this question, then you have eliminated fifteen alternatives. Good work. But if I answer no, then you only get rid of five alternatives. If we assume that I have chosen my number totally at random, then the probability that I will answer yes, and thus eliminate fifteen numbers, is 5/20, since only five numbers give a positive answer. The probability that I will answer no is 15/20, and this answer will only eliminate five alternatives. Combining these two outcomes, we can find the average number of questions eliminated is

$$\frac{5}{20} \times 15 + \frac{15}{20} \times 5 = \frac{150}{20} = 7.5$$

On average, the 'Is it more than fifteen?' question will eliminate 7.5 numbers.

By thinking about the average number of questions eliminated by a question, you can improve your strategy. Think of it like this: when formulating a question 'Is it greater than x?', for which choice of x do we eliminate the most numbers on average?

The answer is $x = 10$, which eliminates

$$\frac{10}{20} \times 10 + \frac{10}{20} \times 10 = \frac{200}{20} = 10$$

questions on average. It is impossible to do better (you are welcome to try), although any question which splits the possible choices into equally sized groups works just as well. We might ask, for example, 'Is the number odd?' or 'Is the last digit of the number between (or equal to) three and seven?' to achieve the same outcome. The advantage of the 'greater than ten?' question is that it has a natural follow-up: 'Is it greater than fifteen?' if the answer to the first question was yes or 'Is it greater than five?' if the answer to the second question was no. The trick to finding the right number as quickly as possible is to keep dividing the numbers into equal-sized groups at each step.

The process of finding a number between one and twenty takes, at most, five steps. The first step eliminates ten numbers, the second step eliminates five, the third step eliminates two (or three) and the fourth or fifth step gives the answer. The tree in figure 16 illustrates the process. The trick to the 'guess a number' problem is to always think in terms of dividing our understanding of a problem into two equally probable scenarios.

The result is that we can break down the problem very quickly. For example, if the number was between one and forty, we would need (at most) one additional question – 'Is it greater than twenty?' – in order to break the options down into two groups of twenty options. In general, we can see the following pattern: guessing a number between one and two requires one question; a number between one and four

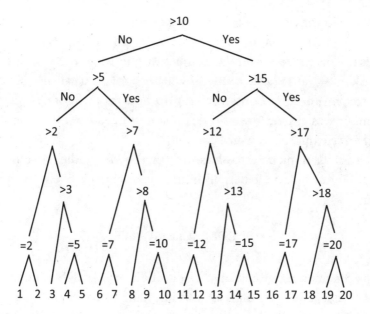

Figure 16: How to determine a number between one and twenty with at most five questions.

requires two questions; a number between one and eight requires three questions; a number between one and sixteen requires four questions. This is because, for one to sixteen, for example, asking four questions allows us to cover $2 \times 2 \times 2 \times 2 = 16$ different options: the number of branches on the tree multiplies by two each time. Continuing building the tree, we find that twenty questions allow us to cover $2 \times 2 \times 2 \times 2 \times 2 \times 2 \times 2 \times 2 \times 2 \times 2 \times 2 \times 2 \times 2 \times 2 \times 2 \times 2 \times 2 \times 2 \times 2 \times 2 = 1,048,576$.

When playing the game of twenty questions, it is then, in principle, possible to identify one out of over 1 million different potential objects, with well-chosen questions. The trick to playing the game is to find equally sized groups for the answer to the two alternatives. So, if you know, for example, that you are guessing an animal, then 'Is it a mammal?' is a good question. 'Is it a platypus?' is not.

The branching tree in figure 16 also helps us to see the connection between twenty questions and entropy and information. To start to understand this link, let's look more closely at the number of questions needed to discover a number. There are twelve numbers which required four questions, and eight which required five. This means that the average number of questions you would need to ask is 4.4, i.e. $(12 \times 4 + 8 \times 5)/20 = 88/20 = 4.4$. You can also see this by tracing the questions from top to bottom in figure 16. If the number was 1, 2 or 3, we needed just four questions. If it was 4 or 5, then five questions were required.

Now let's think back to Shannon's method for finding an efficient encoding in binary. When all letters were equally likely, as in the string

BACDABACDDADBCCB

Betty Shannon suggested a binary encoding of A as 11, B as 10, C as 01 and D as 00. We can think of this encoding in terms of a guessing game. Imagine Claude has chosen one of the four letters of the alphabet at random. The first question Betty would ask is 'Is it one of the first two letters of the alphabet?' If the answer is a yes, she notes down a 1, and now asks a second question: 'Is it the first letter of the alphabet?' If the answer to that is no, she notes down a 0. She now has a sequence of answers yes then no, which in binary is 10, which is the encoding for B. By asking questions in this way, we create a match between yes/no answers and a binary encoding of letters.

We can find a similar encoding for decimal numbers between one and twenty by looking at whether we take the left (no) or right (yes) branch in the tree in figure 16. For example, the number seventeen is yes, yes, no, yes, which when written as a binary string is 1101. Each yes is a 1; each no is a 0. Similarly, the number five is no, no, yes, yes, yes, or 00111. So, the number of questions we need to ask to guess a number is the same as the length of the binary encoding of that number, which is given by the entropy. The entropy of the 'guess a number between one and twenty game' is therefore 4.4 (as we showed above).

In the two examples above, the four letters or the twenty numbers, the outcomes are all equally probable. This isn't always the case. For example, in Betty Shannon's second example, the letter A occurred half of the time, B occurred a quarter of the time and C and D each occurred an eighth of the time. Imagine that Claude again wants to send Betty a message one letter at a time, and she is only allowed to ask 'yes' or 'no' questions about each letter. Now Betty's first question would be 'Is it A?' If the answer is yes, then only one question is required. If the answer is no, then his second question would be 'Is it B?', and if the answer is no, he would ask, 'Is it C?'. If we code a 'yes' answer as 1, and a 'no' as 0, then the letter codes become 1 (yes) for A; 01 (no, yes) for B; 001 (no, no, yes) for C; and 000 (no, no, no) for D. This coding is (again) exactly the one Betty proposed at their dinner and, on average, we will have to ask $7/4$ questions (the entropy) to find an answer. Notice also that there is a direct correspondence between the coding created and the questions we ask. The more likely the letter, the fewer questions we need to ask in order to find out what it is.

In the version of twenty questions where you are trying to identify an object instead of a number the same approach applies. I can't claim to be an expert at the game myself, but several webpages outline questions that split the yes/no outcomes as evenly as possible. Questions such as 'Is it a man-made object?', 'Can you buy it on Amazon?' (after a 'yes' to the first question), 'Is it bigger than a book?', and so on, tend to work well.

The implication of the twenty-questions strategy goes much deeper than the game. We have, in the last few chapters, established that the entropy measures three things: 'the number of questions needed to establish an outcome' (as we have seen here), 'the number of bits required to encode a message about that outcome' (Shannon's theory of information) and 'the randomness of that outcome' (as Esther explained). This relationship tells us that the more uncertain or unpredictable a situation is, the more questions we need to ask before reaching a conclusion.

A good listener always asks questions

To make this last point concrete, let's look more closely at how we should ask each other questions.

Becky's friends often turn to her when they have a problem. They think of her as a good listener, someone who doesn't make judgements about them and takes the time to understand their perspective. Becky enjoys hearing about other people's lives and the challenge of getting to the heart of their problems. In one way, her friends are right. She does take the time to listen. But Becky has a secret, something she doesn't tell the others, about why she is such a popular confidante. It lies less in the amount of time she spends listening, and more in the choice of the questions she poses.

When a friend comes to see her with an issue or something that has upset them, she imagines that there could be millions of reasons for the underlying problem. Her job, as a good pal, is to find out which of the reasons applies to her friend.

Take, for example, the time Jennifer fell out with her colleague at work. One strategy, when Jennifer approaches Becky for help, might be to try to get to the heart of the matter. To ask things like 'Is your colleague an idiot?' or 'Did you fall out because you were late again?' or 'Is it because you had a headache?' or 'Is it because you said something rude?' or 'Did your colleague forget it was your birthday?'

These are not, Becky knows, good starting questions. The answers to these questions will, on average, provide very little information. They are like guessing 'Is it 15?' in the guess-a-number game. Given all the possible reasons Jennifer could have fallen out with her work colleague, and Becky's lack of knowledge of them, the least helpful thing she can do is to start with her own random guess. Not only is it unlikely to be correct, it could very well create tensions.

Instead of going directly for one possibility, and risk being wrong,

Becky starts in the middle. She tries to find the 'Is it more than ten?' question. This can be as simple as asking, 'What went wrong?'; neutral and without assigning blame. Listening to the answer gives Becky a feeling for what happened. She then asks, carefully, 'And how does your colleague feel about it?' in order to get the other side of the story. From there Becky tries to find a new midpoint for the discussion. It may well be that Jennifer is a little more to blame in the situation than she realizes, at which point Becky tries to reposition the chat by finding a new midpoint, a point from which Jennifer might take slightly more responsibility. Maybe the problem now is that the colleague won't accept an apology? Maybe there are deeper issues than the one at hand? Becky remembers, as we saw earlier with Charlie and Aisha, that arguments arise because of the rules of interaction, not because of who started it. This new information – the equivalent to 'Is it greater than fifteen?' – might move the balance back towards the midpoint between the two antagonists. Each question Becky asks, each position she takes, seeks to find a new middle ground, all while still supporting Jennifer. In doing so, she gradually moves closer to a solution to the problem.

Central to Becky's approach is that the less typical a person's problem is, the more questions we have to ask. We see this in Betty Shannon's second example: the letter A, which occurs half of the time, could be identified with only one question. But if the letter was C or D, we needed three questions. It takes more questions, more information, in order to understand people with surprising stories than those with predictable stories. Becky keeps this in mind when talking to people: those with more unusual problems will require her to listen more carefully.

In the classes I teach at the university where I now work there are usually a few students who require extra help and want to come to my office to explain their personal circumstances in more detail. Sometimes, I can be frustrated, feeling that it isn't fair that some students demand more of my time than others.

But then I think about entropy. The university's teaching and my class is built for the typical student. Students with very different backgrounds, the individuals who are more statistically unusual in comparison to their classmates, are also those who contain the most information. It is their stories I should listen to carefully, precisely because they are not typical. The more unusual a person is, the more care we need to take to understand them properly.

Being fair does not mean I allot exactly the same amount of time for every student. It means that I assess each situation using the same steps, eliminating the most common difficulties first. Unusual situations will, necessarily, require more work to deal with. When she is helping her friends, Becky finds herself dedicating more time to more complex issues. If we want to use information fairly, we will spend more of our time dealing with and helping people who don't fit into the norm.

Entropy never decreases

If you have encountered the word 'entropy' in the context of physics, you might have heard that entropy can never decrease or, equivalently, that entropy always increases or remains constant. Inside a water bottle, the H_2O molecules move around, and we quickly lose track of the position of any specific molecule. Once each molecule has had time to travel freely, all molecules are equally likely to be at any place in the bottle. We can see this when we pour oat milk into coffee: at first we know where we have poured the milk in, but over time it spreads out. The position in the coffee of any one oat-milk molecule becomes less and less predictable.

Increase in entropy can be found in all aspects of our own lives. For instance, yesterday I decided to make pancakes. I carefully measured the right amount of flour, milk and egg into a mixing bowl. I poured the mixture into the frying pan, and I'm soon dishing out breakfast to my family. But then I look at the state of the kitchen . . . A random mess! Cooking utensils spread everywhere, flour spilt on the benches, water on the floor. Nothing is where it is supposed to be. This is increased entropy.

To see how entropy increases, let's consider again the number rule we looked at earlier in this section: we think of a number, if it is less than fifty we double it, and if it is more than fifty we take it away from one hundred before doubling. Repeating this process gave a chaotic sequence of numbers. Imagine now that I choose a number and tell you that it is somewhere between 14.0001 and 14.9999. Here I am allowing for decimal numbers, so the starting number could be 14.8538 . It could be 14.1883. Or it could be 14.0016. You don't know the exact value.

Now let's try, through a series of 'yes' and 'no' questions, to find out the resulting number, rounded up to the nearest whole number, after I have applied the doubling rule once. In this case, your first question should be 'Is it twenty-nine?' This is because any initial choice between 14.001 and 14.500 will become 29 when doubled and rounded up (for example, $14.191 \times 2 = 28.382$, which when rounded up is 29); and choices between 14.500 and 14.999 will become 30 (for example, $14.624 \times 2 = 29.248$, which when rounded up is 30). If the answer to 'Is it twenty-nine?' is no, then it must be 30. So, you will only need to ask one question to guess the number correctly – we say that the entropy is 1 after one application of the doubling rule.

Imagine now that I apply the doubling number rule twice, before rounding up. The resulting number will be (after rounding up) between fifty-seven and sixty. It will now take you two questions to find the number. You can achieve this by first asking is 'Is it less than fifty-nine?' If it isn't, then you ask, 'Is it fifty-nine?', which gives us the answer in two. If I apply the rule three times, I will get a number between eighty-one and eighty-eight. Now it takes you three questions. And a fourth time will give a number between twenty-five and forty and it takes four questions.

The number of questions needed increases by one each time we apply the rule. This is what is meant by entropy increasing. As we move forward in time, the number of questions we need to eliminate uncertainty increases too. This is true not only of the number game but also of attendance at bars, Richard's chocolate-cake consumption and the weather. It is true about the lives of married couples after their wedding day (however perfectly planned it was). And it is true whenever I enter a kitchen to make some food. The longer we wait between observations, the harder we have to work in order to find out the current state of affairs.

But then a surprising thing happens. The entropy eventually stops increasing: it reaches an upper limit. Imagine, for example, I had applied the doubling rule described above thirty times to my initial number. You now no longer have any idea what the number is. This is because, after thirty or so iterations, the number which we initially knew was between fourteen and fifteen can take pretty much any value between one and a hundred.

We might be tempted to think that entropy, the number of questions we need to ask to find the number, is thirty. After all, we saw that the entropy increased by one each time we applied the doubling rule. After one step, we need one question on average; after two steps we need two questions; after three steps we need three, and so on. After thirty steps, the entropy should be thirty, shouldn't it?

No. Not at all. Instead, as we saw in the previous chapter, the best way to find out the value now is to start asking 'yes' and 'no' questions for any number between one and a hundred. We first ask, 'Is it greater than fifty?' and use exactly the same steps as we proposed in the previous chapter for guessing a number between one and twenty. By drawing out a tree of guesses (like we did for twenty numbers) we find that the average to identify a number between one and a hundred is 6.72.

There is now no longer any point in trying to keep track of how many times we applied the doubling rule. The entropy is the same for 30 applications as it is for 31, as it is for 100 or 131. Irrespective of how much time has gone, the approach is the same. No matter how many times I apply the doubling rule, the entropy remains 6.72, and the strategy for asking questions remains the same.

This numerical example is somewhat artificial, but the same principle underlies how physicists approach modelling particle interactions. If, for example, we can measure the initial speed and velocity of the white ball in pool when it is hit by the cue, then we can (with reasonable accuracy) predict its position after its first collision with another ball. But any small error in that measurement is multiplied with each collision. Imagine, for instance, if we covered the pockets of the pool table and played thirty rounds of the game. Even if we know, with quite a high degree of accuracy, how hard we hit each shot and the initial positions of all the other balls, it would be extremely difficult after thirty rounds of pool to predict where the white ball will be, even using mathematical models, precisely because the positional error multiplies with each collision.

Engineers, like Margaret Hamilton, might solve this problem by trying to find better ways to measure these systems to more decimal places in order to avoid losing track of the individual billiard balls. They tighten the control. Chaos is avoided by carefully monitoring every detail.

The other way to deal with chaos is to let go. After thirty steps of the doubling rule, or thirty rounds of no-pocket pool, the details become irrelevant. Chaos means that there is no point trying to work out the initial conditions or following the dynamics step by step, or even knowing how many steps have been taken. At this point we should start again, as if we know nothing, and simply ask questions. Is the ball on the left-hand side of the billiard table? Is the billiard ball in the top half of the table? And so on.

The same applies in our own lives. Each sliding door we go through on an underground train, each new person we meet, each decision we make to stop for a coffee or stay inside because it's raining, each word we stumble over slightly as we speak, introduces tiny differences to our lives. As time goes by, the entropy increases. No matter how well we know ourselves today, we cannot know what the future has in store for us.

Living in a distribution

We cannot always live our lives as if we are sitting in mission control. We have to let go. And when we let go, entropy increases.

But letting go produces a new possibility. A way of seeing the world not in terms of certainty but as something blurred out, as a distribution of possible outcomes.

To make this last point concrete, let me describe an experiment I do when I teach the first-year course in statistics at university. The course necessarily requires a lot of theoretical work and long hours in the computer labs fitting statistical models and plotting data. But I believe that the students also need to understand that they themselves are part of a random world. So, when I feel they are ready to have their own existence defined by numbers, I ask them to put down their pens and put on their jackets (it is typically a cold November day in Sweden when I do this) and we all go outside.

I take them to a large quadrangle outside the lecture theatre. I have, in advance of their arrival, drawn eleven parallel lines, about 1.5 metres apart, across the area with chalk, creating ten lanes, like on an athletics racetrack. At the head of each lane, I write a label: 1–3, 4–6, 7–9, up to 28–31, and ask them to stand in the lane corresponding to the day they were born.

The histogram they form is similar to the one shown in figure 17a, known as the uniform distribution, with an equal number of students in each lane. There is some variation: not all the lanes have exactly the same number, and lane 28-31 is slightly different because it contains an extra 'half day' for months which have thirty-one days. But, more or less, the students' days of birth are uniformly distributed.

In one sense, the fact that the students form a similar human histogram every year is hardly surprising: we know the day on which someone is born is random. But what is remarkable is that

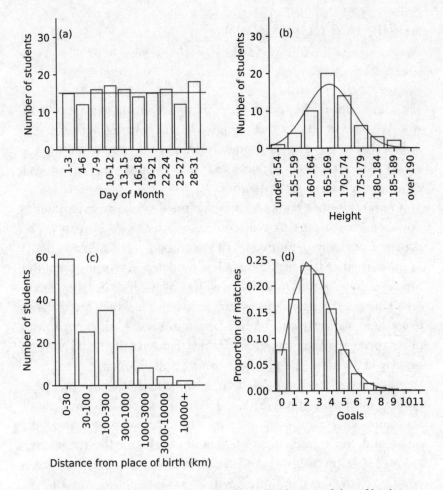

Figure 17: Common distributions. (a) Uniform distribution of day of birth in the month. With the exception of 28-31, each day is equally likely. (b) Heights (in this case of female students) follow a normal, or bell-shaped, distribution. (c) Distance from place of birth for students is a long-tailed distribution. Most students study within 300km of their place of birth, but a small number study more than 3000km or even 10,000km from where they were born. (d) Goals in football matches follow a Poisson distribution. The bars indicate the distribution measured in the data; the solid line is the theoretical distribution, i.e. (a) uniform, (b) normal and (d) Poisson.

I can predict the histogram in advance *because* we know it is random. Think of all the factors that led to the birth of each one of my students. The parents' chance meeting one evening in a bar, or the two long-term friends who fell in love. The drawn-out discussions about when to start a family or the night of passion that led to an unexpected but joyful outcome. The pattern of birth days is very predictable, at the same time as any individual's birthday is completely unpredictable. If all the students in my class claimed they were born on the fourteenth day of the month, this would be surprising, but it wouldn't be random. I would probably conclude that they were joking; taking their revenge on me for making them go and stand in the cold quadrangle on a November morning.

It is when things are truly random that a new way of understanding opens up to us.

Different human characteristics have different frequency distributions. To illustrate this point, I ask the students to stand in the lane corresponding to their height. The sign for the first lane reads 'under 150cm', the second one '150–54cm', and so on, all the way up to '185–190cm' and 'over 190cm'. They shuffle left and right, looking for the correct lane. Slowly, a pattern emerges. The heads of the ten people standing at the front of the lanes become points on a line that slopes from shortest to tallest. But the clearest pattern, from my position standing at an elevated vantage point in front of the students and shown in the photograph which I later show to the class, is in the lengths of each lane. The lane at 'under 150cm' contains just a few girls and the 'over 190cm' contains just a few boys, while in between the numbers in each lane increases. The peak for the girls is in '165–170cm' and the peak for the boys is '180–185cm'.

The typical shape of these curves is captured by the normal distribution, shown in figure 17b. It is characterized by two numbers, the average height of the students (we met averages right at the start of this book) and the standard deviation, which characterizes the width of the distribution, or bell shape. If the students' heights vary a lot, then they form a wide bell. If they vary only a little, then we should use a narrow bell.

Next, I ask them to stand in rows based on how many kilometres they are away from their place of birth. In this case, I divide the lanes into 0–10km, 10–30km, 30–100km, 100–300km, 300–1,000km, 1,000–3,000km, 3,000–10,000km, 10,000+km. The majority are from within 300 kilometres of Uppsala in Sweden, where I teach the course, but there are also those who were born in other parts of Sweden and those who were born much further afield. The distance histogram looks like figure 17c. Most of the students will congregate somewhere around the 100-kilometres median (the midpoint) of the distribution, but there are a small number of people in the tail who come from 5,000 kilometres away, fifty times that distance. This distribution is known as a long-tailed distribution because, unlike the bell-shaped distribution, there are a few people standing in the rarest part of the distribution, its tail to the right. This is very different from the height distribution: the median for female students is 1.67 – I have certainly never had a student who is fifty times taller than average: she would be an 83.5-metre-tall giant!

There is one final distribution that I want to illustrate here but which I have yet to find a way of illustrating in a human histogram. At the beginning of the nineteenth century, Simon Poisson showed that when events happen randomly in time and independently from each other they follow a specific distribution: the Poisson distribution (figure 17d). For example, goals in football occur infrequently during a match and the fact that a team has scored in the seventeenth minute, for example, does not influence the probability that they score in the sixty-fifth (or any other) minute of the match. Goals are both rare and random, and thus Poisson-distributed. The Poisson distribution also captures accidents at workplaces and the number of phone calls you might receive in a day.

After they have participated in the human histogram experiment, I ask my students to create distributions of some aspect of their own lives. Here are just some examples of studies they have created and the distribution they found: prices of apartments in a city suburb (normal), age of chemistry students (long-tailed), points in basketball (normal), membership numbers of student organizations (long-tailed), deaths from alcohol consumption per year in northern Sweden (normal), length of words in different languages (Poisson), length of queue time to use the microwave oven to warm up food at lunchtime (Poisson), number of words in the titles of textbooks (Poisson), arrival times of students to the first lecture of the day (normal), suicides per year in Sweden (normal), heads obtained by coin tosses (normal), ratings of episodes of *Game of Thrones* (normal), season-best times for 100-metre freestyle swimming (normal), bus travel times into the university (normal), number of times the characters speak on the phone in an episode of the TV series *Girls* (Poisson) . . . The list goes on. The fit of the data to the distribution is not always perfect, but it is remarkable just how well the essence of so many different aspects of our lives are captured by these four distributions.

As important as they are in a wide range of applications, I am not going to go deeper into the details of the properties of these distributions. I mention them here to illustrate a broader point: because these four distributions are so familiar to us – we are not surprised when we hear that someone is 175 centimetres tall, was born 1,750 kilometres away, or whose birthday lands on the twenty-second day of a given month – we sometimes forget that these factors are forged from chaos. We saw earlier in this section that we can't expect to control everything, because we don't know everything. We can't measure every butterfly. But once we have understood that limitation and we have let go, a remarkable thing happens. All the resulting randomness produces reliable distributions of outcomes. As a result, our possibility to make predictions returns, in another form. Distributions describe both what is random and what is typical.

Randomness is not something totally unpredictable. Quite the opposite. Randomness comes distributed in a foreseeable way, providing useful models to outline our observations of the world.

Word games

Usually, after Claude Shannon and Betty Moore had eaten dinner, which they now did together every evening, they went back to his apartment. On those evenings, Claude always had to be engaged in some form of intellectual or cultural activity. At first, they would play cards or board games, after which they made music together: her on the piano, him on the clarinet.

Betty soon realized, though, that parlour games alone would not be enough for Claude. He had a need for something more, a creativity at home as well as at work.

So, they started to make up their own word challenges. She would read half a sentence out from a book and challenge him to finish the sentence. These games would often turn words into numbers. He would ask her to guess how many times the word 'the' appeared in a certain page of text: Betty was, like Claude, a mathematician and enjoyed finding patterns just as much as he did.

Betty still thought about their first evening together, when they talked about encoding letters in binary and his scientific article on entropy, but it took some time for her to notice the direction in which their word games now evolved. It clicked one evening when she arrived at his rooms to find that he had prepared the game in advance. He had, he told her, written a paragraph on a piece of paper, and he was going to ask her to guess the sentences he had prepared, one letter of the alphabet at a time. He would note down how many guesses it took her to identify each letter in the sentence. This wasn't a game, she thought, this was a thought experiment!

'T,' she guessed, thinking that the sentence might start with 'The'. She was right. Claude noted down a figure one on the paper in front of him. 'H,' she said, then 'E.' She was right both times and Claude wrote down '1' in both cases.

'It must be a space now,' she said, with some certainty.

'Nope,' laughed Claude, happy to have fooled her with his starting word. 'Try again.'

'Y, then,' she said. 'As in THEY?'

Wrong again. In the end, it took her five guesses to get the right answer: an R. Claude recorded a '5' on his paper. The next letter was easier. It had to be an E, and the first word was 'THERE'. Once she had the first word, the next two were straightforward with only one or two guesses per letter to get

THERE IS NO . . .

Now she was stuck. It could be anything. It took her fifteen guesses to find the first letter of the fourth word, which was an R. The second letter, E, was one guess ('It is the most common letter, after all,' she reminded Claude), but it took seventeen guesses to find the following V.

THERE IS NO REV . . .

From there it got easier, and a few minutes later he had written down the first sentence.

THERE IS NO REVERSE ON A MOTORCYCLE

Once they had completed two sentences, Claude showed her the piece of paper. He had written down each letter in turn, and under them had written the number of guesses she had needed for each letter.

```
T H E R E  I S  N O  R E V E R S E  O N  A  M O T O R C Y C L E
1 1 1 5 5 1 1 2 1 1 2 1 1 1 5 1 17 1 1 1 2 1 3 2 1 2 2 7 1 1 1 1 4 1 1 1 1
A  F R I E N D  O F  M I N E  F O U N D  T H I S  O U T
3 1 8 6 1 3 1 1 1 1 1 1 1 1 1 1 1 6 2 1 1 1 1 1 1 2 1 1 1 1 1 1
R A T H E R  D R A M A T I C A L L Y  T H E  O T H E R  D A Y
4 1 1 1 1 1 1 1 1 1 1 5 1 1 1 1 1 1 1 1 1 1 1 1 1 6 1 1 1 1 1 1 1 1 1 1 1 1
```

'Good guessing,' Claude said. Out of the 101 symbols (letters and spaces), Betty had got it right first time 78 times. The most difficult choice was usually the first letter in each word.

'I think I know why you've asked me to do this,' Betty said. He was, she explained, not testing her ability to guess letters, but instead measuring how predictable the English language is. If English was completely random, then, provided she was allowed to use questions such as 'Does the letter come before N in the alphabet?', she would be able to guess using an average of four or five guesses to get each letter. This was because, she explained, her first question, in this case, would narrow it down to thirteen or fourteen letters, since thirteen letters come before N and thirteen after (if we include a space as the twenty-seventh letter. Her second guess would narrow it down to six or seven letters (for example, 'Does it come before G?'); the third to three or four; the fourth to one or two letters; and the fifth would guarantee an answer (just like the number-guessing strategy we saw earlier).

Claude confirmed her suspicions. He had been thinking about how to measure the entropy of the English language. To understand how much redundancy there is in our communication. The experiment confirmed that if he had been sending the above text as a message to Betty, he would not need to send all the letters. She would be quite capable of guessing many of them for herself. For a communication engineer, such information was invaluable. It meant they could send shorter, more concise messages by telegram.

'If this is how we are going to spend our evenings, then we need to do the experiment properly,' Betty said, walking over to the bookshelf. She took down one of six volumes of *Jefferson the Virginian*, a biography of Thomas Jefferson written by Dumas Malone, which Claude owned.

'Have you read this?' she asked.

Claude confessed that he had not.

'Perfect,' said Betty. 'Let's use it!'

Every evening over the coming weeks, Betty and Claude took turns to choose random 101-letter-long sequences from the Jefferson biography and see how many letter-by-letter guesses it took to decipher each one. They even did the task backwards, guessing a sequence starting with the last letter and then all those which preceded it. Although this latter task felt more difficult for both of them, the number of guesses they required didn't differ much going backwards from going forwards. On average, they calculated, it took about two questions to guess each subsequent letter, provided they knew the preceding eight letters.

It took four or five questions (or bits) to encode letters using the 'Does the letter come before N in the alphabet?' method, which assumes that all letters are equally likely. It took only two questions for Betty to guess the letters using her method, based on her own experience of language.

'This means that about half of what we write is predictable and redundant,' Claude said, 'but half of it remains unpredictable and random. It is here the information lies.'

Taking the high road

John, Richard, Becky and Sofie decide to go to the Cotswolds for the weekend. John, who is driving, checks Google Maps and sees that the quickest route is to follow the M4 motorway. He puts on the GPS and sets off.

Richard has other ideas. A work colleague suggested they take a series of countryside roads through Oxfordshire to avoid the Friday-afternoon traffic. He explains all this to John as they set off, but John has made up his mind. Following the GPS is straightforward, he says, and quicker.

Everything goes well until they are approaching Swindon. After they set off, a lorry broke down and one of the lanes of the motorway is closed. A traffic jam has built up. It looks like it's going to be at least an hour's delay.

'I told you so,' says Richard. 'I said we should have driven via the small country roads.'

'It's easy to say that now,' retorts John. 'I was just following the best information we had at the time! Google Maps couldn't know about the lorry breaking down.'

'But it's just like I said,' Richard follows up. 'You should never trust an app.'

Becky and Sofie, who had been chatting happily in the back of the car, stop talking. The mood darkens.

When something goes wrong, 'I told you so' and 'It was the best option at the time' arguments often flare up, not only for life's small decisions, about which route to take or whether or not we should take an umbrella with us to work, but also for larger decisions in life: moving to a new town or taking a new job. When things don't work out as planned, it is easy to blame those near to us who were involved in the decision.

Chaotic thinking teaches us that there are always going to be large aspects of our world that are random, that cannot be predicted. It is these parts which the entropy measures. It measures lorries breaking down on the M4, leaves on the railway track, a dad putting out his back playing football at a kid's birthday party, a forgotten lunch box, a bike chain coming off on the way to an exam, a big contract falling through at work, or even an unexpected (but not unwanted) new arrival in the family.

Entropy is always with us, created by chaos. We can't predict it and there is nothing we can do about it.

Tempers flare when people attach their egos to the choices they make, despite the fact that no one can really know what will happen in the future. When things go wrong, assigning blame to choices made in the past is just as uncertain an exercise as the making of the decision in the first place. It is seldom true that we 'should have known'. More often than not, we simply couldn't know. Nor should we argue that it was 'the best choice at the time', since there were very likely other equally uncertain choices open to us at that time.

We didn't know then, and we are still none the wiser now.

Instead of wrapping our self-worth in our decisions, we should acknowledge at the time we make our choices that we don't know for sure whether an action will turn out for the best. Such an admission is not a sign of weakness. It doesn't mean that we haven't thought things through. Rather, it comes from an understanding of chaos and uncertainty. It is knowing that much of life is simply beyond our control. We are continually playing guessing games and we cannot know if we will win or lose.

John sits in silence for a while, then he turns to Richard, and says, 'You were right. The way you suggested did turn out to be better. Sorry.'

'That's OK,' replies Richard. 'There was no way to know. There could easily have been a cow loose on the country road that I suggested.'

The mood lightens and, a few minutes later, the lane on the

motorway opens again. The friends' weekend is (probably) going to be a big success.

We have learned about three of the four ways of thinking, providing a structure for dealing with many of our everyday problems. The first stop is statistical thinking: know the numbers. Every fact you can think of – from hand-washing rates after using the toilet (20 per cent worldwide) to the proportion of people who would like to travel in to space (49 per cent of Brits wouldn't go, even if there is no risk) – is only a couple of clicks away. But we also need to realize that the numbers don't usually tell us how we ought to act or how we should interact with others.

It is here that interactive thinking comes in: think through how your actions affect others, and how you let their actions affect you. Understand why you are stuck in a cycle of doing things you don't want to do, or how you end up in pointless arguments, by looking closely at the underlying rules of your behaviour and the behaviour of those close to you. Then, and this is where chaotic thinking comes in, decide which aspects of your life you aim to control closely and which ones you want to let go. You cannot control every aspect of your life, not even a small fraction of it. Instead, get ready for randomness.

When you decide to let go, do so with humility. If you don't know what is happening around you, then ask questions. Don't lose your patience with people who are initially difficult to understand. The more involved a situation is, the more you need to learn. Entropy never decreases. The more time that has passed, the less you know. Remember, you have decided to let go, so there is no prestige to be gained from pretending you know something that you don't. Sometimes you will find the route through the chaos; other times it will be a friend or a colleague who makes the breakthrough. Don't take credit when your luck comes through, and don't resent others when it is their turn to shine.

Looking back, when things haven't gone so well, it is easy to think, 'If only I had followed that piece of advice' or 'If only I had followed my instinct.' When you have made a poor financial decision, been involved in a relationship that hasn't worked out, or made a bad career choice, it is easy to slip into the trap of blaming yourself or others for the decisions made. Of course, you should find out what went wrong and learn from your mistakes. But you should also remember that you never know what the future holds. Don't blame yourself; blame never-decreasing entropy instead.

A sea of words

On the Friday, after the lecture we had at the Santa Fe Institute, I saw Esther sitting alone on a terrace, looking out over the view. The institute sat up on a hill on the outskirts of the town itself and a semi-desert landscape lay before her.

I went out to join her.

We hadn't been alone together since the evening in the library earlier in the week.

'I think I know what you mean about asking twenty questions,' I said.

She continued to look outwards, nodding only slightly in response. So, I continued: we shouldn't assume everyone is average, like Rupert does, nor should we try to claim that there are deterministic and predictable interactions, like Parker does. Chaos inevitably takes over, and we are left with distributions of outcomes. It is these distributions which we need to ask twenty questions about. We should see humanity in terms of our different heights, weights, preferences, ideas, and so on. Everyone is different, but everyone can be understood if we ask enough questions. And populations of people can be understood through the distributions they make up.

In Betty Shannon's example it takes two questions to encode four letters. Twenty questions provide over 1 million different alternatives ($2^{20} = 1,048,576$, to be exact). Thirty questions are enough to describe 1 billion alternatives. Forty questions, and we are at just over 1 trillion. By asking questions and plotting distributions, we can see the outline of the whole of humanity.

Esther sat still for a while and then finally agreed, 'That's quite a good way of putting it.'

Now it was my turn to sit quietly. There was something that still bothered me. I didn't like the way this approach reduced us to

1s and 0s. 'Yes' or 'no' answers. Are we nothing more than points on a distribution?

'What I don't get,' I said eventually, 'is what is the meaning of it all, if we are all just 1s and 0s?'

'Thinking about Lily-Rose and the meaning of life again, are you?' She smiled.

'Aren't you interested in the meaning of life?' I asked.

'I'm Swedish, so no is the answer to that one,' she said, 'The meaning of life is something we study in high school. We are told about each of the world's religions one after the other and then it is explained that there is no one true answer. It is left up to us. We are free to have our own beliefs. And after school we become grown-ups and we don't need to think about such things any more.'

I didn't know what to say. I was unsure if this was some form of Scandinavian humour or a statement of fact.

'The truth is,' she went on to say, 'I got over that when I was fifteen.' She remembered the exact moment. She was sitting like we were sitting right now, but looking out over a different landscape: the sea, at the beach near her family's summerhouse. It was nearing the end of the summer, and she had spent the entire break reading, sitting inside, despite the heat and her parents' attempts to drag her out with them on daytrips.

Her intention had been to read the entire classical literature section of the small local library, in reverse alphabetic order. She had already read and loved Jane Austen, and had a suspicion she might be the best. Not wanting to find every other book a disappointment, she had started from Z to A, instead of A to Z. At first the summer had flown by – Virginia Woolf, Tolstoy, Thomas Mann, Harper Lee, Hemingway, Thomas Hardy, F. Scott Fitzgerald – but then she got to D for Dostoyevsky. Eight or nine thick books, starting with the smaller-scale *Crime and Punishment* and ending with the gigantic tome *The Brothers Karamazov*. The last one had nearly broken her. It seemed to want to provide a deeper meaning, a bringing together of Dostoyevsky's earlier works. It was written with a determination to say something more than anyone else had ever said.

When she finally finished it, she walked down to look at the sea. To clear her head.

As she looked out over the water, she started to wonder if all the words in Dostoyevsky's books were just like the waves on the surface of the water. The more she looked at the waves, the more detail she saw, little variations caused by the wind. But at the same time, it was all the same; the sea just kept moving backwards and forwards. She might not understand every wave, where it was coming from or where it was going, but she understood the sea. It was just the sea. Maybe her relationship to Dostoyevsky was similar. His words washed backwards and forwards over her, in a way that was varied and rich, but in the end, everything in the book was the same. All she had done was spent the summer reading words. Nothing more, nothing less.

She thought of the arrangement of all the molecules of water in the sea, and then thought about all the possible arrangements of words she had and hadn't read. She didn't need to continue reading, she thought, because it would just be more of the same: the typical words that a typical great author writes in typically great books. She now had a feeling for literature, just like she had a feeling for the sea. From there on she could just enjoy rereading Austen, safe in the knowledge that the sea would always be there.

It was a few years after that, when she went to university to study computer science, that she learned about representing data in binary. Her lecturers had told her that everything was bits of information: waves, both in the sea and in the light; and words, both written and spoken. They showed her how to most efficiently encode those words, to turn them in to bits. This allowed them to be sent all over the world, as little packets of light, just like the ocean currents transported water. It was, as she had felt that day she stared out at the waves, all bits: 1s and 0s. Nothing else.

She had learned about entropy and randomness. She had researched the uniform distribution, the normal distribution, long-tailed distributions, Poisson distributions and many other statistical techniques.

What she had realized, something that she felt her own supervisor, Parker, had not fully grasped, was that these distributions solved many complex problems.

'I mean, Rupert does need to catch up on a few of these techniques,' she said, 'but when he criticizes Parker's overall approach, he has a point.'

She was, she told me, annoyed by the way Parker presented predator–prey models, chaos and dynamics: as if there were something more in it than there actually was. He was, she felt, making out that there was something magical going on, but she understood it best when she just focused on the mathematics and the statistics.

It was exactly the same as Dostoyevsky. So many unnecessary words . . .

'Ironically,' she said, 'Parker doesn't seem to get the central message from entropy: that we should find the information in the message and strip away the noise. He pads out the message with his own brand of nonsense.'

She had realized that there was an alternative. It came when she started studying Claude Shannon's articles. She was particularly fascinated by the study he had conducted, together with his wife, Betty, measuring the entropy of language. They had suggested how to predict the next letter or word in a sequence using the words and letters that came before. Recently, when Dostoyevsky's works were made available online, she had done exactly this. She found that, using Shannon's theory, she could predict Dostoyevsky's next word in a sentence from those that came previously. Not with 100 per cent certainty, of course, but there was a clear underlying structure to his prose. Even the most complex texts fell into a predictable distribution.

The idea of focusing on data was central to something we had heard was being discussed at Stanford: a new idea about data mining. There were these two guys, postdoctoral researchers, Sergey Brin and Larry Page they were called, who were planning to teach a course in the fall term on this approach. She had finished her masters with Parker, and although he had offered her a PhD position, she had decided to move to California and learn from Brin and Page instead.

Think about the World Wide Web, she said, all the things you can search for when you open Mozilla. But instead of focusing on what is actually written on the pages, think instead in terms of the links between the sites: which page connects to which? Which of the sites is the most popular? It doesn't matter what you read on a particular website, all that matters is that some pages are more connected than others. If we can find the distribution of the popularity of all sites, then we can help people find the information they need.

In the future, Esther said, we will have massive amounts of information about individuals: what they read online, what TV programmes they watch, what they bought at the supermarket, who they were friends with. Everything. We can think of the choices people make on the internet as a stream of 1s and 0s. Each click of the mouse is a decision of what they want and what they don't want. We will be able to predict those clicks and design algorithms that automatically identify the information they want; the products they might buy . . .

'Isn't there anything more?' I asked. 'Surely we can't just see people in terms of how often they visit different internet sites? Just like you shouldn't see Dostoyevsky as a sea of words. He has something meaningful to say.'

'Maybe, maybe not,' she said, 'but that isn't my point.'

She was, she explained, when thinking about work, concerned only with things that we can measure and predict. We are scientists, she told me, we measure things. We measure the pressure and temperature of the sea, the size of the waves. She found that the frequency of different words in Dostoyevsky's writing followed a long-tailed distribution, just like the number of connections between webpages. The structure of the internet could be understood in terms of distributions and entropy.

'Your friend Lily-Rose is, of course, free to talk about the stars and our minds and all that, but Parker can't. Not here, anyway. Not at work,' she said. 'When we can measure something, then we should.

There are patterns in everything. When we find them, we can talk about them and we can use them to help others find the information they are looking for. But when there is no pattern, we have nothing to say. We should remain silent.'

We sat in silence and looked out on the landscape, an evenly spaced dotting of bushes on the hillside: a low-entropy arrangement of shrubbery.

Everything seemed to stand still and, in that moment, I felt I understood exactly what Esther meant when she talked about the sea. There was a stillness in seeing the world as encoded only in 1s and 0s. A calmness in describing the world as distributions, entropy and likelihoods.

Could there really be anything more? Or had I reached the end of my journey?

Class IV: Complex Thinking

The World Congress

Andrej Nikolayevich Kolmogorov stood and waited in front of the blackboard. He was nervous, a feeling he was not accustomed to. When in front of the small classes of PhD students he taught at his summerhouse over intense weekends of study, or in front of pupils and groups of teachers in the schools he visited, he felt in complete control. Back in the Soviet Union, Kolmogorov was admired by everyone. Despite having few political connections, he had risen through the academic ranks to such a degree that this trip to Nice, France, in 1970, which would not have been sanctioned for many of his comrades, was immediately approved for him. His intellect, and his achievements, had made him untouchable.

But this stage, and this large audience, was different. It was the International Congress of Mathematics, the meeting of elite mathematicians at which the prestigious Fields Medal was awarded once every four years. Kolmogorov knew that the audience contained many of the members of the 'Bourbaki' group. These mainly Paris-based mathematicians were puritans of the subject. Under the pseudonym Nicolas Bourbaki, they had collectively authored a series of textbooks called the *Elements*. The project aimed at establishing an approach to mathematics based on the greatest possible rigour. They believed that their approach was to be not only the basis for all research but also the way the subject should be taught from the very earliest years of school.

It was this group's very palpable presence that put Kolmogorov out of his comfort zone. The Bourbakis' gaze was focused on him alone. They seemed to carry the stern demand that what he now said should conform to the rigour they had set up collectively. It was worse than the Politburo, he thought. If he considered it rationally, he knew he needn't be worried. After all, forty years earlier, as a young man, he had provided the first framework for probability theory. This had proved a key part of the Bourbaki *Elements* project. He had then gone on to solve so many important mathematical problems that, at one time, a rumour had spread in French mathematics that Andrej Kolmogorov was not just one man but was the collective pseudonym, equivalent to Nicolas Bourbaki, for Russian mathematicians. But he was, of course, just one man. A very nervous man.

He wanted to tell the Bourbakis the same things he told his students, the schoolteachers and the wide-eyed twelve-year-olds who he would meet in his teaching: to discuss with them those small problems, the challenges he saw as lying at the border between the trivial and the impossible, which interested him. Tasks which even a schoolchild might have a chance of solving but could at the same time completely stump even the most experienced professors. He wanted to then talk about why grand projects, like their *Elements* textbook, that tried to unify the entire field of mathematics were always doomed to failure.

Finally, he started his presentation.

'I wish to begin with some arguments that go beyond the framework of the basic subject of my talk,' he said, hoping to prepare his audience for what was to come. And, as he said these words, he realized that there was no way back, that he had to go through with it. He had to say it exactly the way he saw it.

'Pure mathematics is the science of the infinite,' he pronounced, his voice growing bolder, 'and Hilbert, the founder of the conception of completely formalized mathematics, undertook his titanic work merely to secure for the mathematicians the right of staying in paradise.'

There was an audible gasp from the audience. David Hilbert's 'titanic' project was the famous set of twenty-three mathematical problems that had been announced seventy years earlier at the 1900 International Congress of Mathematics, which, once resolved, were meant to position mathematics as the only way of reasoning rigorously. Hilbert wanted mathematics to become the science of the infinite, all-encompassing, all-explaining. But now, Kolmogorov, some of whose earlier work had been central to the construction of Hilbert's project, was implying that the grand mathematical ship was always doomed to sink.

He outlined, as an example of the difficulty Hilbert's project was facing, the way in which Bourbaki had defined the number one in their *Elements* textbooks. For Kolmogorov, and for every school-child, the number one was the number one. There was nothing complicated about it. But when Bourbaki had defined the number one, they had started by defining a way of reasoning based on Venn diagrams, illustrating relationships between sets of objects. Only once they had defined sets of objects, over many pages of symbolic manipulations, could they define the number one.

To Kolmogorov, their way of working was misguided. He told them that all their complicated calculations went against what he saw as the way a schoolchild does mathematics: using their intuition. Under the Bourbaki approach, a schoolchild was meant to be able to understand the formal world of Venn diagrams before they understood the fact that one person, one cow or one dollar could exist! Defining the number one should, of course, be straightforward. A complicated definition must be the wrong one.

Bourbaki mathematics had failed at the most basic level, he said. It was making simplicity complicated and failing to address true complexity. Kolmogorov said the iceberg that had now hit Hilbert's 'titanic' came in the form of the computer. In the future, he explained, problems in physics and other sciences would not be formulated in terms of the infinite mathematics offered by set theory and Venn diagrams but coded directly in the form of computer simulations. Today, in the Soviet Union and the USA alike, engineers build simulations of everything from spacecraft flight to the economy. He declared that they were moving toward the era of the algorithm: sets of instructions which told students how to find the right answer and computer code that reliably implemented our instructions.

Kolmogorov realized he had stated enough generalities. After all, they were here to do maths, not to talk about it in mixed metaphors of sunken ships, infinite paradises, or ways to get to Hilbert's heaven. And he suddenly understood why he had felt so nervous before he had begun: he could only follow up such an introduction with a truly novel mathematical result, something his audience had never heard before. And in that moment, looking down at the notes in his hand, he knew that this was exactly what he had.

As his nerves disappeared, he looked up at the audience and pronounced loudly, 'By thinking in terms of algorithms, I am now going to shed new light on a fundamental mathematical problem: what does it mean to say something is complex?'

The Bourbaki collectively leaned forward in concentration, silent. They would not be convinced by imagery of how Soviet computing might replace their belief in an infinite mathematical paradise. But they might be convinced by logical argument. Now it was time to listen and consider . . .

The matrix

'What Esther is describing,' Max told me, 'is the matrix.'

Over dinner on Saturday, I had told Max about my conversation with Esther at the Institute the day before.

Max explained to me that at a fundamental level, she sees the universe as many computer scientists do: as a vast array of 1s and 0s. It is this view he had come to call the matrix. It was, after all, the mathematical name for an array of numbers. But it was also an evocative word, one that captured the unfathomably large nature of data in the modern world.

When Esther looks at the matrix, she asks which parts of it are possible to predict. She knows there is randomness, but she aims to reduce the dimensions of the randomness to something she can understand. She looks for the pattern.

'I have no doubt,' Max said, 'that people like Esther will become increasingly powerful over the coming decades.'

He reminded me of the first evening we spent in the sports bar, with all the noise from the sports channels. He told me that the world will become increasingly dominated by that type of noise, not just in American bars but all over the planet. Artificial versions of sports, online games, like the computer game *Doom*, will become more realistic and all-consuming. Our heads will be eaten by virtual-reality headsets. We will start to connect with each other like never before. It will no longer just be academics and nerds using the World Wide Web but everyone, chattering, debating, sharing pictures and sounds, all the time, everywhere, all at once. Our concentration levels will drop, as we flip from one activity to another, unable to discern between the important and the trivial. News, sport, politics, gaming, opinions, facts and fiction will all combine to one almost infinite source of entropy.

'Those people who can sort and organize that information will become rich and successful,' Max went on. He told me that Esther and her soon-to-be colleagues at Stanford will mine the matrix to find the things that the masses find most entertaining. Then their algorithms will create more artificial versions of the text, music and pictures, causing the matrix to grow exponentially. He anticipated that those who just stare into space as the information expands, like Lily-Rose stares out at the night sky, will lose out. They will find more and more to wonder at but also lose track of the difference between reality and illusion.

'Even those with the power to extract information from the matrix will lose their ability to see the truth,' he said, 'but it won't matter, because reality itself will change form.'

'So, Esther is right?' I asked. 'Everything is just information and probability distributions?'

Max looked at me. His monologues were usually delivered to a point slightly below my right shoulder, but now his gaze fixed directly on me. I felt compelled to look away.

'Have you read Shannon's paper properly?' he asked, pointedly.

I had to admit I hadn't really had time to take in all the details . . .

'The reason we are here, or at least the reason I am here,' he proclaimed, 'is that we reject the way Esther sees it. She is wrong. You and me, we refuse to see the world purely in terms of randomness, just as we refuse to see it as purely linear or stable.'

If I had read the original entropy paper in the way he had, I would have known that Claude Shannon was clear that his work had nothing to do with true complexity. In the introduction, he wrote dryly that 'frequently the messages have meaning; that is they refer to or are correlated according to some system with certain physical or conceptual entities'.

Max told me that this was an understatement. 'Shannon is telling us that his theory deals with almost none of the things we think of as important, like the things that surround us in the physical world and definitely not the world of concepts and ideas. It has nothing to do with true complexity.'

Shannon wrote in his article that 'semantic aspects of communication are irrelevant to the engineering problem'. By this he didn't want to imply that the meaning (semantics) of messages was unimportant. On the contrary, he wanted to emphasize that his approach did not deal with the things that were *most* important. His entropy was only a technical solution to help with storage and transmission of information. It didn't tell us what the information we received in messages meant to us.

Take music as an example, Max said. In 1949, the year she married Claude, Betty Shannon worked on an algorithm to automatically generate music scores. Together with her colleague at Bell Labs, John Pierce, she devised a system of rolling dice and following mathematical steps that produced chord progressions. Mozart and Bach had, in their time, also used randomness to produce music, but Shannon and Pierce's work formalized this procedure in equations and improved the way in which chords were constructed. The results were mixed. Some of the pieces sounded 'rather musical', according to the algorithm's creators, but they also admitted that there was a lack of connection between the chords and a clinginess or jumpiness which meant the songs didn't quite work.

'That is the problem with music generated by a computer,' Max said. 'It always misses something. A depth. An emotion. A meaning.'

After they married, Betty and Claude Shannon started a family and moved to Boston, where Claude was made a professor at MIT. They continued to work on technology-based projects. Claude designed a robotic mouse that solved mazes, and Betty completed its wiring. They jointly devised a scheme for stock market investment which saw them become successful early investors in rapidly growing Silicon Valley firms. But neither of them took a genuine interest in entropy or abstract measures of information. They focused instead on practical activities that had real meaning to them as individuals.

It was this point that Esther had missed, Max told me. She reduces everything to probability distributions, he said. But this leaves us with an approach that is like Betty Shannon's early algorithmic music: disconnected chords, jumping between ideas, songs that don't quite work. We cannot capture the nature of humans with these methods. We can't get to the heart of our complexity.

'So, you agree with Lily-Rose?' I asked.

'No!' he exclaimed. 'Of course we don't accept that pile of astrological mumbo-jumbo.'

Max told me that when people like Lily-Rose are confronted with the vastness of the matrix they let their gaze shift out of focus. They see only the mystical.

'We need to be more critical in the modern world, not less so,' he said.

He told me that what we – he, Chris and (he hoped) me – were looking for was the true theory of complexity. A theory that sat somewhere between randomness and chaos on the one side and order and stability on the other. A theory that took into account interactions but moved past the simpler examples of predator–prey cycles and susceptible-infected-recovered models of disease.

Such a theory, he told me, would need a change of perspective. It would need us to acknowledge that we live in a world of a trillion dimensions and to make a determined effort to find a new path through them.

Max sounded now less certain. 'That is why we are here, isn't it? That is what we are trying to find out. We all want to know the true nature of the matrix. Isn't that the most important question of them all? What lies in that massive array of 1s and 0s? Where can we, as individuals, be found in all this data? What do we truly understand of physical and conceptual entities, as Shannon put it?'

'But what is that theory?' I asked.

'Well, I suppose that is what we are going to hear about during this final week from Chris. He is going to tell us the secret of complexity . . . Or at least he is going to tell us where all the bright minds working here have got to in their search for it . . .'

Up until now, Max had seemed to have an answer to every question I posed, but now I realized that I couldn't push this particular point any further. He didn't know the secret to this sort of complexity. The question for both of us now was what that secret might be.

Four people in a car

We started our journey by finding out that statistical methods – the mean and the median, maximum likelihood, straight-line relationships in data – were useful for looking at patterns in society but not powerful enough to capture many of the things that are important to us personally. This led us to look more closely at our interactions – the rules underlying our discussions (and disagreements), our social epidemics and the tipping points we experience. Then we discovered chaos. Randomness is unavoidable and often arises because of the extreme measures we use to try to get back on track – crash diets or drastic resolutions. It is then we look again at the numbers. Not averages this time, but distributions of outcomes – heights, wealth and our personal histories. Chaotic thinking taught us to let go. But to ask informed questions in order to find out who others really are.

The success of these three ways of thinking lies in breaking down individual problems: understanding why traffic jams are beyond our control; identifying situations where saving time can make us happier; looking closely at how we respond to others; thinking about which aspects of our lives we should control and which we should let go of.

And yet in many of life's trickier situations, there is an extra dimension, a different level, something that we can't simplify or break down. Take, for example, John, Richard, Becky and Sofie sitting in a car together, soon to arrive in the Cotswolds. While we might be able to help John and Richard to stop trying to alpha-dog each other over their map-reading skills, we have not addressed any of their deeper issues. Maybe Richard was irritable because he is anxious about problems at work; maybe John really wants to impress Sofie on their weekend away; maybe Becky is annoyed with

Sofie because she is only there because John fancies her; maybe Sofie is not in the least bit interested in John and is instead thinking about the long runs she can make through the countryside . . .

When the four friends got into the car, they took with them their history, their relationships and their innermost thoughts. These are not reducible to a small set of well-defined relationships.

To put it simply: their lives – all of our lives – are complicated.

We can't necessarily break this complexity down. That is exactly why it is complicated.

But what we can do, and this is where Andrej Kolmogorov comes in, is find a definition: we can find a way of measuring how complex something is.

Only as complex as its shortest description

The difficulty in defining complexity, Kolmogorov realized, lay in saying precisely what it is that makes one thing more complicated than another. Is a river network flowing out of a mountainous region more complex than a straight canal cut through the country-side? Is the turbulent flow of air created at the tip of an aeroplane wing more complex than the ripples created by a boat moving slowly through water? Is the flip of a coin more complex than the falling of an apple from a tree?

In some cases, these questions can appear to be Zen-like riddles. From the perspective of Google Earth, the network of streams down the mountainside are more complex than a countryside canal, and yet the canal was the result of human ingenuity, of intricate minds and elaborate relationships which are themselves much more complex than mountain geography.

Kolmogorov provided a one-sentence answer to complexity rid-dles in his presentation in Nice in 1970. He stated that *a pattern is as complex as the length of the shortest description that can be used to produce it.*

This is why the canal itself, which can be described as 'a straight line cut out of the ground between A and B', is less complex than a river network, which would require a description of the contours of the mountainside. But Kolmogorov's definition also explains why the process of planning and building a canal – involving the coord-ination of workers, the manufacture of complex tools, engineering principles and division of labour – is more complex than the process by which a river network is formed – as a result of water slowly pushing through earth, stones and sand.

Kolmogorov's answer links our ability to describe something suc-cinctly to its complexity. Although the flip of a coin is chaotic, a mathematical description of its trajectory is similar to that of an apple falling from a tree, albeit with the addition of an equation to describe the coin's rotation. So, using Kolmogorov's definition, a coin toss is only slightly more complex than an apple falling to the ground. Likewise, the apparently random sequence generated by the doubling rule in the last chapter is not complex, because we could capture it in a single equation. Nor is air or water turbulence complex, because it is generated from a rather simple process of an object moving through a fluid.

Kolmogorov's genius, and I believe his definition of complexity to be one of the twentieth century's most important and under-rated discoveries, was to see that complexity depends on how good we are at explaining it. Before Newton proposed his theory of gravity, other theories might ascribe to every object its own unique place in the world: apples should fall to the ground in October, the moon should go around the Earth, a human's place is on the ground while a bird's is in the sky. By providing an explanation in the form of the theory of gravity, Newton replaced innumerable complicated explanations with a small series of concise mathematical equations to describe the motion of objects – which would accurately describe a wide range of future observations too.

Without the people trying to do the explaining, nothing is complex or simple. Indeed, one of the ways to describe science is that it is the process of finding progressively shorter explanations for the phenomena around us. As scientists find these explanations, apparently complex things suddenly become simple. It is the things that are hard to explain which are complex.

This was a very different view than that held by other mathematicians at the time. At the start of the twentieth century, one of the key problems in mathematics posed by David Hilbert was to define axioms for probability theory. Axioms are statements which are considered self-evident, ones that no one could reasonably question. In 1933, Kolmogorov proposed three axioms for probability: (1) events cannot have negative probability; (2) at least one event must have a probability of 100 per cent; and (3) if two or more events are mutually exclusive (they cannot both occur), then the probability that at least one of them occurs is the sum of the probability of each of them occurring.

To make these more concrete, let's consider the example of a six-sided die. Axiom 1 says the probability we get a six cannot be less than zero (however the die is formed). Axiom 2 says the probability that we get a number between one and six when we roll the die is 100 per cent (assuming the die has six sides, labelled one to six, and that it can't land on its edge). Axiom 3 says that the probability we get either a five or a six (which is 2 in 6 for a fair die) is the probability of getting a five (1 in 6 for a fair die) plus the probability of getting a six (also 1 in 6). These axioms are, I think we can all agree, beyond reasonable doubt. What Kolmogorov did, to solve Hilbert's problem, was to show that all other reasonable propositions about probability followed from these three axioms alone. For example, they could be used to work out the probability of getting two sixes in a row when throwing a die, or the probability of throwing a die ten times and never getting a six. Everything we know about dice (and probability in general) follows from these three axioms.

Kolmogorov's axioms, which he discovered in the 1930s, were aesthetically pleasing to Hilbert and the Bourbaki group of pure mathematicians and were widely celebrated by his peers. But, by 1970 Kolmogorov saw his axioms as being too abstract. If we wanted to explain dice to a child, we would not start by telling them that the probability that a die lands on any particular side cannot be negative (as axiom 1 tells us). Because this piece of information is obvious, it is also pretty useless when given as part of an explanation. We might talk instead about how, because the die bounces around a lot, it is hard to predict how it will land. This latter description is the algorithmic approach that underlay Kolmogorov's new approach in 1970. It also underlay a scepticism, which he voiced in his presentation in Nice, of the entire programme of Bourbaki-style mathematics. As he said then, '[in Bourbaki mathematics] the meaning of the "number 1" contains some tens of thousands of signs, but this does not make the concept of the "number 1" inaccessible to our intuitive understanding'.

This example, of tens of thousands of signs being used to define the number one, was just one of many examples of how an insistence on using what might appear to be the simplest form of mathematics, in the form of axioms, leads to overly complicated explanations of the real world. It was why Kolmogorov abandoned the axiom-based approach and started to think in terms of information and computer code – things which express finite descriptions of how we see phenomena in the real world.

We have already seen, in Part III and in Max's image of the matrix, that data can be written in strings of binary. For example, all words and text can be encoded as 1s and 0s using the ASCII code (the eight-bit character code used in modern computers) or in terms of the yes/no answers to questions. The images on the screen of your phone can be encoded in terms of pixels, themselves binary strings, describing the intensity of red, blue and green at every point on the screen.

Algorithms for making computations can also be written in binary. There are many different programming languages – Python, C, JavaScript – but all of them are translated into binary code inside the computer's processor. Whenever we write a piece of computer code, it can be represented as a string of 1s and 0s.

Kolmogorov defined the complexity of a pattern as the length of the shortest algorithm that can be used to produce that pattern. For example, a program to paint a computer screen completely white is short. It would loop over all the pixels and set every value to 0 (assuming 0 is white). A program to draw a straight line is also short, specifying the start and end coordinates of the line. The same is true of a program to draw a circle or a square. So, according to Kolmogorov's definition, white screens, lines, circles and squares are all simple.

More complex patterns, the graphics in a computer game for example, require longer code and are therefore more complex. Games like *Tetris* or *Wordle* that are visually simple can be written using a small amount of computer code. Games with more complex graphics, such as *Fortnite* or *Grand Theft Auto V*, require much longer computer programs to run them.

Kolmogorov's insight, made long before we had computer games, was that complexity was not a property of the output itself. Complexity is the length of the program that generates or describes the output.

The streets of London

We don't use sets of binary rules, computer programs or algorithms to communicate with other humans, but we can still learn from Kolmogorov when describing others.

To see how, let's follow Aisha's work for a charity that helps homeless people in London. The statistics are shocking: one in fifty-two people living in the capital are homeless. That is over 170,000 people. But somehow, Aisha finds that when she tells these numbers to others (including her closest friends), they still don't grasp the scale of the problem. Even when she presents the statistics to an audience of government decision-makers or charitable donors, she often feels they switch off and stop listening.

The problem, Aisha thinks, is that they can't grasp the different dimensions of the problem. It isn't just about those unfortunate souls lying in city-centre shop doorways. The true extent of homelessness is much larger. Many homeless people are stuck in hostels, sofa-surfing between houses or squatting illegally. And their problems vary. Aisha is aware of how difficult it is for someone who doesn't know where they will be sleeping from one day to the next to hold down a job, to build a stable relationship, bring up children and maintain their mental health. She witnesses the tragedy every day in her work. She has experienced their lives, felt their troubles and understands the challenges they face.

In terms of Kolmogorov's complexity, saying that 170,000 people in London are homeless is not enough. It is a short explanation, but it is too short. Statistics are valuable, but one number can't capture the complexity of all these people's lives. For Aisha, focusing on numbers has not proven a successful strategy when trying to persuade policy-makers and potential funders to take the problem more seriously.

Frustrated by her lack of success using numbers, Aisha tries a different approach. The next time she is asked to give a presentation about her organization to policy-makers, she asks one of the women she has worked with, Jackie, to tell her story for them.

Jackie's difficulties had come without warning. She had a good job and a steady income, and she was enjoying life and travelling the world. But when she lost her job she found that she couldn't pay her rent and got heavily into debt. Evicted from her flat, she moved from place to place, relying on friends and acquaintances, all her belongings packed in her car. She became depressed and, although prescribed antidepressants, only six months after she had lost her job took an overdose. She wanted to end it all, and she nearly did. It was only then that she made a decision to refocus and find the positives in her life. The coming year would be different, she said to herself. With help from Aisha's organization, Jackie has found a place to stay and got a temporary job. Her belongings are still in storage, and she remains in debt, but she can now see a path through her difficulties. The future is bright.

After Jackie speaks, Aisha's audience ask how they can help other people to improve their circumstances, as Jackie has; to find the fortitude to make changes; to take that next step. Aisha tries to explain that each experience of homelessness is unique, that personal fortitude is not always enough – external help, advice and care are often large factors in helping homeless people. In Jackie's case, it was the help she received after the overdose that turned things around. For others, different help is needed. It might be Aisha or one of her colleagues talking to an unfriendly landlord and drawing up an arrears agreement that makes the difference or help with alcoholism; giving someone a chance at work, or anti-depressive drugs, or the opportunity to talk to a counsellor. Aisha talks about feedback cycles, where people become depressed and lose their home then enter a deeper depression from which they struggle to escape. People become isolated, more so when they have nowhere to stay or have to move elsewhere to find accommodation. And segregation is a problem: immigrants find it harder to access services and have no one who can point them in the right direction.

But whatever Aisha says, now the policy-makers have met Jackie, their focus is on her. How, they ask, can they help others follow Jackie's path, to make the change she did?

Aisha is frustrated. Jackie has told her story incredibly well, but hers is just one story out of 170,000. After the presentation, potential funders seem to think that homelessness is solved by motivating people to find their own way out of their troubles, just like Jackie did. That is not what Aisha intended at all! A single story hasn't built an overall case for why Aisha and her colleagues' services are so important to one in fifty-two Londoners.

Each person has their own challenges, and it is impossible to communicate everyone's story in a fifteen-minute presentation. She wants to get across the scale of the problem, that it is happening to a wide range of people and that each individual needs help in a different way. How can Aisha convey the breadth of experiences of homelessness?

It is then that the insight comes. Her job is to plant a seed of each of the many homelessness stories in the heart of her audience. From there, inside the hearts of others, the complexity of their situations should grow, just as it did when they heard Jackie's story.

Aisha selects the seeds from the varied lives of the homeless people she has met and worked with: those who have lost their job, fallen into alcoholism, whose relationships have broken down, or who have returned from travelling or service in the armed forces to find the life they once had no longer there. She thinks about how to tell each of these stories in a way that blossoms into a shared understanding: a concise but accurate overall picture of the problem.

Aisha selects three more individual stories: an insurance broker who turned to alcohol when his girlfriend left him; a refugee arriving from Syria with no connections or access to help; and a man who had been sleeping on the street for twenty years and had given up hope. Each of the four personal tales captures part of the experience of being homeless and, with a small TV production company, Aisha makes a video depicting their lives, telling their stories. The camera doesn't linger on any one person too long and the film moves in and out of the four lives, occasionally panning out to show the city as a whole to emphasize the scale of the problem. The video conveys both individual elements and the overall pattern.

Aisha's new approach is the essence of Kolmogorov complexity. The more our audience hears about a particular person, the more stuck they become in the details of that story. But we cannot expect others to create their own internal experience if the only seed they are given is numbers. The secret then, to capturing complexity, is to find relatable stories that are both personal and capture variety, then let them grow within the intended audience. We don't need to tell every detail of the story we want to convey.

I, II, III, IV

It was the Monday morning of the last week of the summer school. Parker stood in front of us and wrote three bold Latin letters on the blackboard – 'I, II, III' – then smiled at Chris, next to him, who wrote 'IV' to finish the sequence.

We all knew something special was coming. Parker was handing over to Chris, who would tell us the secrets Max had talked about. Even Alex, who had skipped most of Parker's lectures (because he had 'more interesting things to do') had come to this lecture on time. By him sat Max, hardly able to hold back his excitement, and further down the line was Rupert, who was trying to hide his. Madeleine and Antônio, who had become inseparable over the last few days, sat in the row behind.

I sat in the middle of the lecture theatre next to Esther. Zamya sat two seats along from us. She had been quietly taking notes in all the lectures, annotating everything with coloured pens. Today was no exception. I could see the same I, II, III and IV written at the top of a page in her notebook.

Chris explained that the roman numerals stood for four classes proposed by Stephen Wolfram, the first person to thoroughly investigate the elementary cellular automata models we had worked with in the computer lab. Wolfram had hypothesized that elementary cellular automata could produce four different types of behaviours: (I) stable, (II) periodic, (III) chaotic or (IV) complex.

We had already seen numerous examples of classes I, II and III in our lectures. Chris reminded us of his examples in the computer lab a couple of weeks earlier. The domino rally of 1s moving from left to right converting 0s along the way, and the Democrats (0s) and Republicans (1s) forming groups on either coast were class I patterns:

stable and unchanging. The chequerboard of 1s and 0s Esther had created (figure 9) was periodic, class II, as was the fractal shape I had made in the computer lab (figure 13). The pattern produced by the other cellular automata was random (figure 14) and with help from Esther I had shown that it produced unpredictable shapes and was thus class III. Parker added that he had shown us many examples of class I and class II systems: predator–prey cycles, tipping points and the outcome of a social epidemic. He had also started our journey into chaos, class III, by showing us the Lorenz system.

Chris said he was now going to show us a specific elementary cellular automaton with rules that can be written as follows:

111	110	101	100	011	010	001	000
0	1	1	0	1	1	1	0

As before, these rules show how the three closest cells in the row above determine the cells in the row below in the same way as I studied them in the computer lab.

'This rule looks much like the others,' he said, 'but as we will now see, it is very special.'

Chris showed us an animation: starting with just one black (1) cell at the top in a row of whites (0), the rows successively filled the screen. Unlike the cellular automata we had looked at in the computer lab, which had spread both left and right, this spread only to the left. A mess of small triangles moved outwards, leaving behind more regular, wave-like patterns, which were themselves broken by another mess of small triangles (figure 18a).

Chris took us on a zoological tour of the cellular automata. Depending on how it was initially set up, it would create vertical lines, messy triangle structures that moved at differing speeds, solid lines which moved from one side to the other, and many other weird and wonderful shapes that interacted on a sea of much smaller triangles (some of these are shown in figure 18b). When two structures met they morphed into another structure, which would move across the sea to meet yet another structure and morph once again. The

231

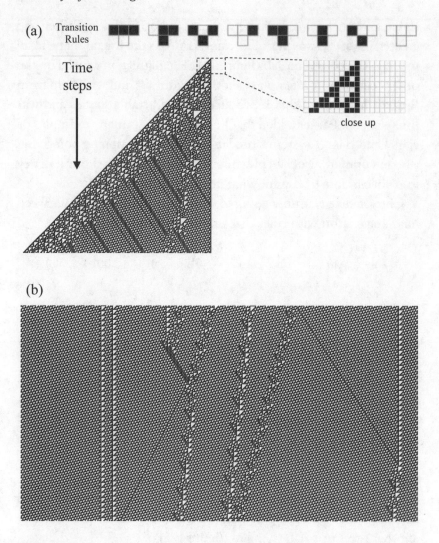

Figure 18: (a) An elementary cellular automata that produces a complex pattern. (b) Some of the structures seen within the elementary cellular automata.

outcomes in the slides were just some examples, Chris told us, of an endless zoology of combinations.

The rule Chris showed us now was neither regular, periodic or random. It was complex. It was class IV.

The lines and squiggly structures are known as *emergent* patterns,

Chris said. The original rules of interaction – the rules by which the three nearest neighbours determine whether the cell in the next row is black or white – are simple, but these squiggly structures had a life of their own, seemingly unrelated to the original rules. These complex structures emerge out of the simple, local interaction rules, Chris explained.

Chris told us that Wolfram had hypothesized that every process, biological or physical, personal or social, natural or artificial, lies in one of only four classes of behaviour he had observed in his computer simulations of elementary cellular automata. It was the final class, Class IV, Wolfram believed, that was the most important.

Discovering and describing emergence in elementary cellular automata was, Chris claimed, the missing link in understanding complexity. Kolmogorov had argued that a system was only as complex as the rules that generated it, but he didn't have a sufficiently powerful computer to allow him to investigate the relationship between the rules and the patterns. The Russian was therefore unaware that there were examples of phenomena that looked complex but were in fact (according to his definition) simple. This was exactly what Wolfram did. He investigated how different sets of rules led to different patterns. Then he documented the rich zoology of structures which cellular automata could produce.

Wolfram hypothesized, Chris told us, that the whole of life itself could be the outcome of cellular automata-like rules. Biological life, with all its twisting, turning, dynamic structures, might be nothing more than the output of a simple updating rule. One which we just don't know yet. Even our own minds and consciousness might have emerged out of such simplicity.

'Do you believe that!?!' shouted Antônio, sounding excited by the idea. 'Do you think that a rainforest is just the outcome of one of your simple, computerized rules?'

'I can't say I totally buy the theory,' smiled Chris.

But, he went on, it appears that the complex patterns in elementary cellular automata lie on the edge of total randomness and order. The other day in the lab, he said, David had shown him two examples of elementary cellular automata (figures 13 and 14 in Part III). One built a fractal-like shape, a recurring branching structure. The other produced a completely random pattern. The patterns we see in the complex cellular automata lie on the border between these two. Complexity is created at the edge of chaos and order, he said.

Antônio loved this idea. He said that it aligned with his own thinking when he worked deep within the rainforest. When he thought about the branching of the Amazon River in the basin across Brazil, he saw it feeding and being fed by the vegetation. The rainforest itself was a wild tangle of different layers, plant growing on plant, insects feeding on vegetation, mites parasitizing insects, and micro-organisms swarming through everything. He told us that when he was deep within the forest he felt that there was an essence, a form of simplicity that summed it all up. This was what had motivated him to become a theoretical biologist, he said, to try to find the formula which described that feeling he had.

For once, even Madeleine didn't try to stop Antônio talking. Chris was listening intently. He had, while Antônio spoke, started up the cellular automata simulation again. It scrolled upwards on the screen, filling it with new and different patterns generated by this new rule.

When Antônio had finished, Chris summarized. The output from a cellular automaton isn't exactly the same as the rainforest Antônio described, he told us, but it is its own jungle. Complex patterns, he said, can emerge from the simplest of rules.

All of the life

Wolfram's elementary rules were not the first cellular automata (CA). The idea of setting up a grid of cells then updating them was first proposed by Stanislaw Ulam and John von Neuman in the 1940s. But cellular automata research came to life, so to speak, in the 1970s, when Cambridge mathematician John Conway proposed what he called 'the Game of Life'. Conway's cellular automata operates on a two-dimensional grid (as opposed to the one-dimensional line of cells used by Wolfram). Every cell is in one of two states, alive (black) or dead (white), and on each time-step it updates by looking at the state of the eight nearest neighbours and applying the following rules

1. A living cell with only one live neighbour dies (black to white).
2. A living cell with two or three living neighbours remains alive (black to black).
3. A living cell with four or more living neighbours dies (black to white).
4. A dead cell with exactly three living neighbours becomes a live cell (white to black).
5. A dead cell with any other number of live neighbours than three remains dead (white to white).

We can think of rule 1 as death by loneliness (not enough neighbours), rule 3 as a death by overcrowding (too many neighbours) and rule 4 as reproduction by the three surrounding cells (it takes three cells to have a child cell in the Game of Life!). I am quite sure there is no real biological system that reproduces in exactly this way, but the model captures something of the essence of life: isolation, overcrowding and reproduction.

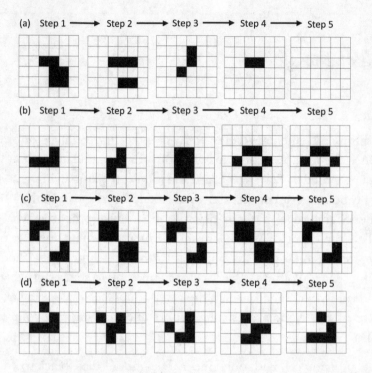

Figure 19: Illustration of the Game of Life. Structures on a six-by-six grid which (a) die out, (b) reach stability in a 'beehive', (c) oscillate periodically and (d) move across the grid in a 'glider'.

The outcome of these rules applied to four six-by-six grids of cells is illustrated in figure 19. In the first example (figure 19a), we start with a healthy-looking clump of six cells on the first step. On the second step, the middle two cells are overcrowded (more than three neighbours) and die, and one new living cell is added on the top right of the structure (because it has exactly three neighbours). Now, on the third step, the bottom pair of cells does not have enough neighbours to survive (rule 1, loneliness), while the top three cells change form somewhat. On step four, one further cell is lost; on step five, the remaining pair dies. The final, stable configuration is rather solemn: everything is dead.

In the example in figure 19b, there are only four cells to start with, but it grows to a solid block of six cells (on step three) before

Step 1 ──────────────────► Step 2

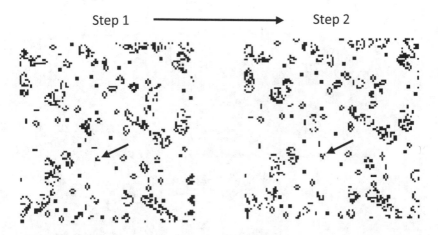

Figure 20: Illustration of the Game of Life. On the 100-by-100 grid in a wide variety of different patterns form.

stabilizing as what CA enthusiasts call a beehive: a stable arrangement of cells in a squashed hexagon. Periodic, oscillating patterns also emerge, as in, for example, figure 19c. In this case, the two triads of cells each has a child who, in the next generation, dies from overcrowding, allowing them to have another child on the next step. The cycle continues indefinitely.

The structure shown in figure 19d is one of the most important in the Game of Life. It is called a glider because over a sequence of four steps it changes form and moves down and to the right of the array of cells. These gliders will, unless they meet another shape, continue to move in the same direction.

Figure 20 shows two consecutive steps of the Game of Life on a 100-by-100 array of cells. Several stable beehives have formed, along with many other shapes, both stable and unstable. It is this rich diversity of forms that gives the 'game' its name of 'life'. The arrow in figure 20 points to a glider travelling slowly down to the left across this landscape which will eventually crash into a stable square block further down towards the bottom-right-hand side of the array.

There are many complex structures to be found in the Game of Life. The top part of figure 21a illustrates an example of such a

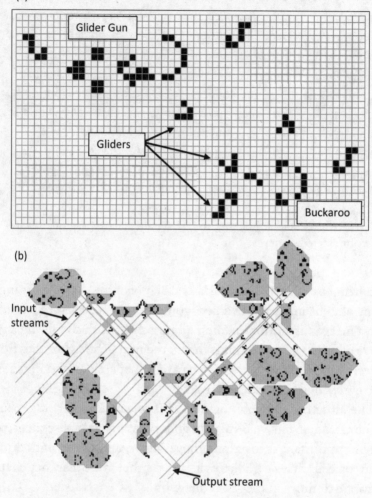

(a)

Glider Gun

Gliders

Buckaroo

(b)

Input streams

Output stream

Figure 21: More complex structures in the Game of Life. (a) A glider gun oscillates backwards and forwards to create a stream of gliders towards the bottom right, where a Buckaroo deflects them and sends them to the bottom left. (b) David Buckingham and Mark Niemiec addition machine. Two input streams of gliders come in from the left. Black and white indicate the state of the cells; grey-shaded areas illustrate the form of the larger structures. This cellular automata adds together the inputs and will output the sum in the lower output stream. These drawings are adapted from Paul Rendell's PhD thesis, *Turing Machine Universality of the Game of Life*, University of the West of England, 2014.

construction, known as the glider gun. The gun oscillates backwards and forwards, emitting a new glider every thirty steps; these travel off into space. The gliders emitted can be sent as input to other structures. For example, the structure at the bottom of figure 21a, fondly referred to by cellular automata specialists as a Buckaroo, takes in a glider and reflects it at a ninety-degree angle.

The Game of Life produces dynamic emergent patterns on a scale much larger than the original rules had. The 36-pixel-wide and 9-pixel-high glider gun is already larger than the local interactions between the cells (on a three-by-three-pixel grid). Combining Guns with Buckaroos and other shapes – given names like Queen Bee, Fanout, Pentadecathlon Reflector, Splitters and the Takeout by CA enthusiasts – it is possible to produce even larger dynamic structures.

These CA fans have gone on to build calculators and computers using Guns, Buckaroos and the other shapes. In the 1980s, David Buckingham and Mark Niemiec showed how to put together about fifty of these shapes to build an adding device which takes in two streams of gliders and outputs their sum (shown in figure 21b). The gliders are used to shuttle information backwards and forwards between the various shapes. Another researcher, Paul Rendell, built a full-scale computer using cellular automata.

Game of Life has only two possible states ('alive' and 'dead'), but cellular automata can, in general, have more states. With some basic programming skills, you can build your own. One masters student on a course I teach about modelling complex systems, Mikael Hansson, created a CA he called the Labyrinth Factory. The rules for the cells – which I call bone/white, goo/black and fluid/grey – are as follows:

1. Bone cells continue to be bone if they have four or more bone neighbours (white to white); otherwise they become goo (white to black).
2. Goo cells remain goo while three or more of their neighbours are bone; otherwise they become fluid (black to grey).

3. Fluid cells become bone if two or more of their eight neighbours are bone (grey to white); otherwise they remain fluid.

One resulting labyrinth, built after 3,000 steps of the automata, is shown in figure 22. White bone walls, two or three cells thick, have formed. Each of these is surrounded by a single-cell layer of black goo, and within these walls there is a mixture of white, grey and black cells. Within this mixture, bone turns to goo, goo turns to fluid and fluid turns to bone again. The outcome is a sequence of dynamics waves which oscillate and interact to form chaotic and complex patterns.

When I look at the Game of Life or Michael's Labyrinth Factory and then I look out of the window, at the trees blowing in the breeze

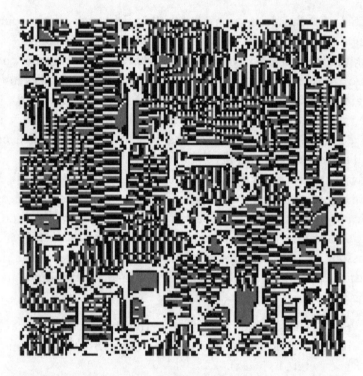

Figure 22: The Labyrinth Factory. Snapshot of the outcome of a cellular automaton with three states (white, grey and black) following the rules described in the text.

and the birds flitting between the branches, I can't help but feel a connection between simulation and reality. Both the computer and nature reveal complex patterns of movement. Nature is, of course, deeper than the Labyrinth. The tree has roots that spread far underground and is built of innumerable cells transporting minute nutrients throughout its structure. The birds' bodies contain complex organs performing vital functions and brains processing and reacting to many different sources of information each moment. But the cellular automata suggest that at least some of the complexity that we see in the natural world might be attributable to simple rules of interaction. Could a similar secret be held within nature? Is nature, as Wolfram hypothesized, built on this sort of simplicity?

This last question remains unanswered. Many scientists would contend that the question doesn't even belong in the category of true scientific investigation. On most days of the week, when I am not looking at the Labyrinth or out on a beautiful summer day, I would probably agree with them. Wolfram's hypothesis is too vague. But it was a question similar to Wolfram's that first interested polymath John von Neumann in the world of cellular automata. He made his curiosity more concrete by setting himself a challenge, to find a self-reproducing automata: a system that produces children which then go on to have their own children. Self-reproduction, von Neumann believed, was the hallmark of biological life, and finding it inside a computer would help explain the roots of biological organization.

Santa Fe researcher Chris Langton* partially solved von Neumann's problem by building what he called a self-reproducing loop. Langton's cellular automaton has eight states, and each cell updates itself based on the state of its four nearest neighbours (up, down, left and right) and its current state, using a total of 219 rules (there are five rules in the Game of Life). The initial form of Langton's loop is shown in figure 23 (step 0). Think of it as a sort of worm.

* Langton is part of the inspiration for the character Chris in Santa Fe, who I met while at the summer school in 1997.

The 2s can be thought of as a 'skin' surrounding a 'core' of 1s. The 7 0 and 4 0 pairs placed within the core cells are something like a genetic code. They propagate down through the worm, telling it which way to turn as it reproduces to form a new worm.

The process of loop reproduction is shown in figure 23. The loop grows larger between 70 and 120 steps and at step 151 the first loop has had a child-loop. Both these loops go on to produce further children. The original loop will now produce a loop above its current position while the child will produce a new child to the right. Over time the population of loops will grow until the computer screen on which this cellular automaton is run has filled with a spiral of loops (bottom of figure 23).

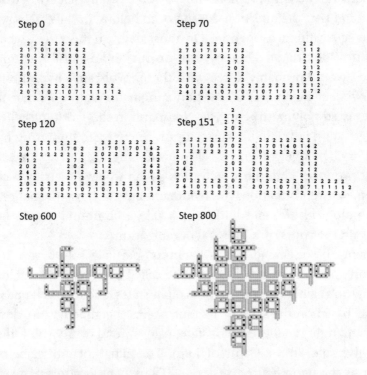

Figure 23: The Langton Loop cellular automaton. The initial loop is shown in step 0. Each number represents a state of a cellular automaton (as described in the main text). For steps 70, 120 and 151 the states are numbered, while in steps 600 and 800 the cells are coloured from white (state 0) to black (state 7).

Langton coined the term 'Artificial Life' to describe his investigations into how various aspects of biological life can be reproduced inside a computer simulation. There have been a wide range of different approaches to Artificial Life – including letting little computer programs compete with each other for computer memory, building artificial chemistries in which numbers 'react' with each other to build larger, more complex numbers and creating far more advanced Langton loops (some of which even have loop sex with each other) – but it remains a research field filled with unsolved questions. We can see complex shapes emerging from computer simulations and we know that biology is full of complex structures. But, so far, scientists haven't been able to find a rigorous connection between these two observations. We can't create true artificial life inside a computer.

There are, though, many ways in which understanding the emergence of global patterns from individual interactions become an important part of technology and science. When filmmakers want to create artificial landscapes, whether it's forests and mountains here on Earth or alien environments on science-fiction planets, they use three-dimensional fractals. With a few lines of computer code, they can produce endless and varying patterns that mimic the contours of the natural (and supernatural) world.

Some amateur enthusiasts have embraced the spirit of Kolmogorov, by competing to create the most complex of landscapes from the least possible code. One of them, a Twitter user called @zozuar, composes Tweets of computer code shorter than the character limit of 280 characters but which generate lifelike videos. Figure 24 shows some snapshots of his work: trees, mountainous forests, clouds, waves, ancient and modern cities – all have a Kolmogorov complexity of less than 280 characters.

The approach of describing local rules of interaction to help explain emergent patterns has during the last twenty-five years become an important part of every area of science. My own research has shown how we can reproduce the motion of animal groups using what are known as self-propelled particles. Each simulated animal (particle) interacts with nearby particles through simple

Figure 24: Some examples of complex patterns that can be generated by less than 280 characters of code. Work by @zozuar on Twitter.

rules of attraction, alignment and repulsion. These models can be used to reproduce the twisting and turning of a starling murmuration on a summer evening, the flocking of sheep as they are herded over a field, the flash escape of a school of mackerel when a shark attacks, and locusts swarming the Sahara Desert. Other researchers use a similar approach to describing the growth of cancerous tumours, embryo development, the growth of plants, the firing of neurons, and many other biological systems.

Finding models that explain how the components of a system interact to build a global pattern is part of the bottom-up approach. We saw this approach when we looked at interactive thinking, with two types of individuals: rabbits and foxes; infectives and susceptibles; ants that know where the food is and ants that don't; dominoes standing or fallen; Democrats and Republicans; people shouting or not shouting. And these individuals interact in relatively simple ways. They influence each other, they spread rumours, they make each other happy or sad, more or less motivated to exercise.

Now we see that much more complex patterns can emerge from these bottom-up interactions. The interactions of the birds, the fish, the branches of a tree, the cells or other units that make up a system are described, and the outcome at the global level emerges from those interactions. By describing systems from the bottom up in terms of local interactions between the individual units we can explain more complex patterns at a higher level. A bottom-up description of individual birds helps explain the apparent complexity of the movement of the flock. A bottom-up description of cancerous cells helps explain the growth of a tumour.

The hard edges of social reality

This bottom-up approach hasn't just found widespread application in biology, it is also key to explaining our social lives. We are all part of a bottom-up system. Each of us follows our individual rules of interaction and out of that emerges the complexity of our society.

To help us investigate this social emergence, let's follow Jennifer to her university's main library. Earlier this year she enrolled for a masters course. She felt she needed a change. She wants to learn more, change her view on life and, ultimately, earn more money. But this meant leaving her friends in London to move to study up north, in a city where she doesn't know anyone.

The reading room she is studying in now was built almost 150 years ago. It is spacious, the ceiling three storeys high, and walkways allow access to the books, stacked all the way up to the ceiling. The room contains seven rows of desks, and each row contains thirteen desks. At each desk there is one chair. In total, there is seating for ninety-one people.

In the reading room the rule is total silence. The high ceilings and hard wooden surfaces of the desks and walkways make the slightest sound reverberate throughout the room. Every noise Jennifer makes – from opening the door, walking in, finding an empty desk, drawing out the chair, sitting down and taking out her laptop and books – is an excruciating scraping of surfaces. She feels the disapproval of those around her who are trying to study and have been disturbed by her arrival.

For Jennifer, the advantage of studying in the reading room is that it is a commitment. She knows she is going to spend the next few hours here, because she also knows that to leave, or even to take her phone out of her bag, will make too much noise. She is frozen

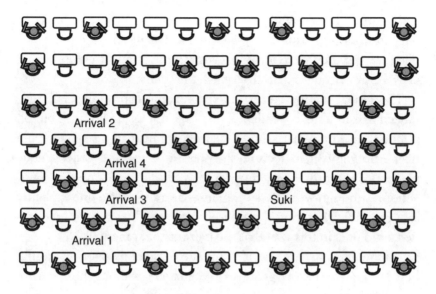

Figure 25: Arrangement of students sitting in a library. A chequerboard is formed as students avoid sitting next to each other.

in fear of disturbing someone else, and so is everyone else. And in that frozen state, all they can do is work and study.

This frozen state has a very particular form. Both desks next to Jennifer are empty, while those further away are occupied. In her row, six people are seated, at least one desk between them. The desk behind her is empty, but the two desks next to that are occupied. The same is true in the row in front of her, where Jennifer has people sitting diagonally to the left and right of her. If we were to look down from above, we would see an imperfect chequerboard of desk occupations, where most of her fellow students have empty desks next to them (figure 25).

Let's now consider the bottom-up, individual rules that generated this imperfect chequerboard. The first people to enter can sit wherever they want. Often the back of the room is slightly more popular than the front and the seats there will fill up first. But at this early stage people generally sit anywhere in the room. It is these initial placements that lead to the imperfection of the chequerboard. For example, arrival

2 in figure 25 came slightly after arrival 1. Arrival 3 chose to sit diagonally in front of arrival 1, so that when arrival 4 came the room had filled up and they were forced to compromise and sit either behind arrival 2 or in front of arrival 3 (which they chose to do).

The global seating pattern in the reading room, a socially distanced frozen state, is familiar to all of us, not just in libraries but on public transport (what kind of weirdo sits next to a stranger on a near-empty bus!), in lecture theatres, in cafés and other public places. It is also an emergent property of our social rules of interaction. Although we avoid sitting next to strangers, no one explicitly said that library seating should look like a slightly deformed chequerboard. It is in this way that the pattern that emerges.

Jennifer has started to notice similar emergent patterns all around her. When she passes groups of teenagers walking home from

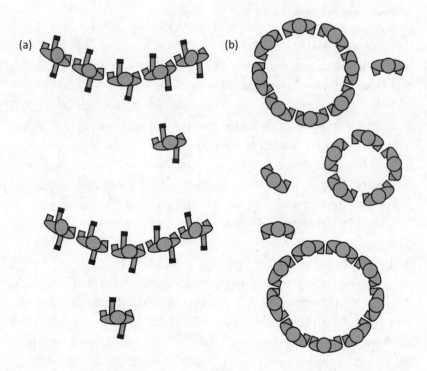

Figure 26: (a) Schoolkids walking home along a path. (b) Closed circles of friends.

school at 4 p.m., they form V-shapes where, in groups of five people, the teenager in the centre is furthest back from the direction of travel (Figure 26a). These structures enable the kids to talk to each other; by angling their bodies inwards, everyone can be part of the conversation. Groups of three to five are relatively stable, but groups of seven take up too much space on the pavement. It is also difficult for those on the outside of a large group to talk to each other. As a result, in some cases, one or two individuals have fallen away from the group, walking on their own. In other, larger groups, the kids jostle to try to get to the middle, before breaking up and re-forming in smaller groups one behind the other.

The schoolkid who finds herself on the edge of a V-shape has to work hardest to hear what everyone else is saying. It is now her responsibility to deal with the group's potential collision with eld-erly passers-by going in the opposite direction. Feeling detached from the group, the girl furthest out eventually drops off and walks on her own. She looks at the girl in the centre of the group, every-one facing inwards to listen to her, and her own loneliness is compounded.

When we enter a library reading room, we have all agreed that we want the space to study in peace. But Jennifer has also felt a simi-lar distance in her lectures, like that of the kids who fell away from the group as they walk home from school. When Jennifer sits in the lecture theatre, she always leaves an empty chair, or even two, between herself and her fellow students. Other students do the same, leaving space for her. Jennifer wants to move closer, but the way she sees others sitting creates a distance which she can't seem to reach through. Outside the lecture theatre, during breaks, she notices other students standing in a closed circle of friends, while she is left standing alone (Figure 26b). By standing in a tight circle, the members create the shortest possible distance between each other. But Jennifer is left outside.

When we look at the collective pattern – a lecture theatre full of people sitting seats apart, a circle of friends, a V-formation of schoolkids – we get the impression that the pattern reflects how we

249

as individuals want the world to be. Often, it doesn't. It can feel like the loneliest thing in the world to fall off the end of a group on the walk home from school. But it wasn't (necessarily) the intention of the kids to push that individual out.

Structures such as the V-shape might well reflect the social hierarchy of teenagers: the most popular kid is probably in the middle, the second and third most popular are to her left and right, and so on. But the structure that emerges as we walk together exaggerates and reinforces that hierarchy: those on the outside have to work harder to be part of the conversation than those in the middle and have the highest risk of dropping off the edges. The V-shaped structure amplifies the insecurities of those on its edges. These physical structures we create – often for reasons of ease or efficiency – can form social boundaries that often feel much more entrenched than the people who are part of them intended.

At Christmas, Jennifer returns to London for a large party, attended by some of her friends. Imagine one hundred people standing in nine groups are chatting over drinks before dinner. In total there are sixty men and forty women. Now imagine John, who is standing in a group with two other men and seven women (Group A in figure 27 step 1). They are talking about the *Sex and the City* sequel, *And Just Like That*. John is a bit bored so he politely excuses himself and walks off to join another group, one with four men and three women, who are talking about football, a subject he thinks he knows a fair bit about (Group D). Sofie is in a different group (Group C), one made up of two thirds men. They are talking about Bitcoin and NFTs – not her favourite topic, so she joins another group nearby (Group F) that has a more even balance of men and women. Both John and Sofie's moves are illustrated in Step 1 of figure 27.

John and Sofie's decisions change the gender balance of the groups they leave and the ones they join. Sofie's arrival in Group F, along with several other women who have decided to leave other male-heavy groups, now means that more than two thirds of Group F are women. The guys in this group are starting to feel a bit

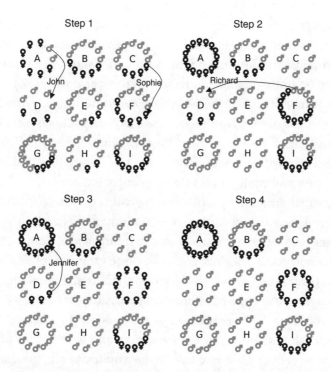

Figure 27: A model for an office party. Nine groups made up of a total of sixty men and forty women are standing in groups. Men who are in groups made up of more than two thirds women leave to join groups where the majority are men. Similarly, women who are in groups made up of more than two thirds men leave to join groups where the majority are women. The steps show how the group composition changes over time. The names relate to people described in the text.

outnumbered and wander off to join another group. One of them is Richard, who joins John's new group, as illustrated in Step 2 of figure 27. The arrival of Richard and one other person in Group D means that it contains six men and three women. Now, Jennifer, who was in group D; she didn't come all this way to talk about Arsenal's winning form. She decides to leave and join group A, where they are talking about her favourite new TV series. Jennifer's move is illustrated in Step 3. By the end of this process of moving around, five groups are men only, two are women only and only two groups have mixed membership (Step 4).

Although I have described the movements of people in figure 27 as if they are individuals, their movements are decided by a mathematical model under which they all follow the same rule. Specifically, individuals in a group where a third or less of people are the same sex as they are will move to a randomly chosen group where the majority have the same sex as them. This is a relatively weak preference on the parts of the individual – everyone is happy to be in a minority – yet after everyone has moved, the majority of people are in all-male or all-female groups.

At first, when Jennifer looks at the groups that have formed, she is tempted to conclude that men and women prefer not to talk to each other. But then she thinks more carefully about how their local interactions produce a global outcome. She knows her friends don't want to form gender-segregated groups: this is just the pattern that emerges from each individual person's decision. It would, for example, with some advance planning, be possible to construct nine groups, each containing six or seven men and four or five women, in which all groups have at least one third men and one third women and in which everyone is happy with where they are. Jennifer thinks about how she might do things differently. She could, for example, grab three female friends and join a group of men. That way a good balance can be maintained within the group and everyone will feel included.

The way social structures emerge from individual interactions lets us cut through the complexity. Patterns at a group level or within society more broadly should be seen not on their own terms but in terms of the actions of the individuals who create them. The individuals are often behaving in a much simpler (sometimes even unconscious or unthinking) way than it appears when we look at the structure as a whole. Using his definition, Kolmogorov would say that the social hierarchy of teenagers or the web of social networking at London parties are not as complex as we first thought. Once we have established the rules of interactions, we have reduced the apparent complexity to an understandable simplicity.

This lesson carries with it a responsibility, for all of us. If you are an influential or popular person – either because of your role at work or through your social status – think about how you can use your physical location to avoid excluding others. Don't form closed circles of friends which prevent others from joining. Look behind you when walking with others and check if someone is left walking on their own. Now and again, sit next to someone you don't know in a lecture and exchange a few words with them. The way we interact, inadvertently and collectively, creates hard edges between us. It is your individual responsibility to see where they lie and to soften them.

A person is a person through other people

What makes our social lives even more complicated is that the rules we use change, depending on the social outcomes we experience. Complexity becomes layered upon complexity.

In the same way that the Taoist idea of yin and yang is a useful way of seeing the distinction between order and chaos, the oral tradition Ubuntu, known in many parts of Africa, is a useful way of seeing these many layers of complexity. The Ubuntu worldview can be summarized as the idea that 'a person is a person through other people'. It is a humanist philosophy, best known in the West as one of the guiding principles in the post-apartheid Truth and Reconciliation Commission in South Africa. In Archbishop Desmond Tutu's speech, 'No Future without Forgiveness', he said of Ubuntu that

> I am fully me only if you are all you can be. Anger, resentment, nursing grudges corrode, subvert the summum bonum, the great good of the African worldview of communal harmony and they eat away at the very vitals. To forgive is not being altruistic; it is the best form of self-interest. You know what happens to your blood pressure when you are caught in a traffic jam, 'How come they let all those morons drive a car?' To forgive is [as] good for your physical health as it is for your spiritual health.

In Tutu's example, the traffic jam is an emergent structure, arising from all our desires to get home after work. Apartheid is also an emergent structure, segregating people along lines of race. That is not to say that being stuck in apartheid is the same as being stuck in a traffic jam, but it is to recognize that, within these examples, we cannot clearly separate the individuals from the social structure.

While Tutu emphasizes forgiveness and harmony, Ubuntu is more than this. It is a deep understanding that we are defined by our inter-actions with others. In the previous chapter we saw that we can't look at a group of men talking at a party and conclude that they just want to talk to other men, or that nobody likes the kid walking behind the group, or that we should never sit next to someone we don't know at a lecture. But not everyone participating in the struc-tures see it that way: a man who enjoys talking about football starts to search out all-male groups; a girl who is left out on the walk home starts to define herself as a loner; students who can't sit still in a library (wrongly) conclude that studying isn't for them; and the person stuck in a traffic jam starts to see all other drivers as idiots. Ubuntu emphasizes that we are shaped as individuals by the systems we find ourselves in. We are people, through other people.

We become most aware of how our social environment defines us when we are in crowds. At a Fatboy Slim beachside concert in July 2002 researchers studied how some of the 65,000 people attend-ing the event experienced it. The crowd – three or four times more people than the organizers were expecting – were packed tightly together on a narrow stretch of uneven beach between the incom-ing tide and the stage. Security personnel were unable to make their way in among the crowd, and to many of those looking on from the outside the situation appeared dangerous and uncertain. But the concert-goers saw it differently. When surveyed afterwards, those in the most densely packed areas reported that they felt less crowded there and gave a more positive account of the event than those on the outside. A sense of belonging and social safety emerges from their collective closeness.

If we view people purely in terms of their physical existence, then packing them together is *not* safe. At densities of seven or more people per square metre, then the crowd can become a fluid mass, with shock waves potentially carrying people over long distances. At the Hajj pilgrimage – where 2 and sometimes 3 million Muslims make their way to Mecca each year – densities can reach up to twelve people per square metre at peak times. This has led to sev-eral disasters. In January 2006, 363 pilgrims died in a crush on the

Jamaraat Bridge. Nine years later, despite rebuilding parts of the route, disaster struck again and thousands more lost their lives.

Yet, like at the Fatboy Slim concert, the Hajj attendees who identified strongly with others on the pilgrimage felt safest at highest densities. This feeling is explained by a perception among individuals that the rest of the crowd are supportive of them. Researchers describe a 'virtuous circle' among dedicated pilgrims. They search out the highest densities, where other participants also identify strongly with the event. This in turn leads them to feel a sense of wellbeing, surrounded by others they identify as 'good Muslims', which then enhances their belief in the importance of the pilgrimage.

Physical interactions within crowds can already, in themselves, generate complex patterns: from the small-scale V-shaped groups of pedestrians (figure 26a) to large-scale pressure waves at pilgrimages and concerts. But people are more complex than billiard balls. Individuals in crowds change the bottom-up rules that they follow, precisely because of how they feel inside the crowd. So, the emergent pattern at a Fatboy Slim gig and on a pilgrimage to Mecca is not just a physical mass of moving bodies but also a new sense of social identity, a connection to others.

Muslims who have attended the Hajj have been found to have a greater belief in harmony among religions and are more peacefully inclined. Attendance at Fatboy Slim concerts hasn't been followed up with questionnaires, but it is easy to imagine, years later, how two acquaintances who find out that they were both there relive their experiences. Being at the centre of the crowd, crammed together in a way that is potentially dangerous, provides a shared feeling that stays with us for the rest of our lives.

It is shared social identity which brings people physically closer and creates the positive emotions they feel in crowds. In a study by Anne Templeton conducted at Sussex University, she gave one class of around 120 second-year psychology students a baseball cap with a 'Sussex Psychology' logo on it to remind them of their shared identity on the course. She and her colleagues then filmed the students as they walked from the lecture theatre along a path to their next activity. When the students had caps on, they walked more closely together, more slowly and in larger groups than they had when leaving the same lecture the week before without caps. Instead of walking in groups of two to four, and with a few people walking on their own, they now took up the whole walkway in larger groups of six or seven. Their social identity changed their interactions.

The reason for gathering can vary – from religious or educational reasons to partying – but people who identify with each other want to be closer to each other and are happier.

Next time you are in a crowd, take some time to reflect. Think about the physical interactions with those near to you. Think about the simple ways in which you follow others, move closer to and away from others. Think about the group feeling within the crowd. Think about the effect your crowd has on those who haven't joined it and on those who have formed other groups. Think about how the groups you join and leave are shaped by society and how they shape and change society. Think about the history that formed your group and how it might shape history in the future. Most of all, realize that through all of these levels of social interaction *you* become a person through other people.

Here it is!

During the last week of the summer school, Chris had shown us many of the ways in which mathematical models were used to help understand the biological and social world. He emphasized that in our research, when creating mathematical models of reality our aim should be to find simple rules which produced emergent patterns. He reminded us that no model is perfect so, whenever we create a computer simulation or model which explains a natural phenomenon, we should remember that it would always leave many aspects of reality unexplained.

'But that's okay,' he told us. 'Our job, as scientists, is to use models to help us. We might never come to a complete understanding of the complexity of our existence. But what you should do is find short explanations of how the parts in the system interact. That way, you are getting to the essence of complexity.'

Chris's final lecture on the Friday was different. He closed the laptop on which he had been showing us computer simulations and started sketching on the blackboard. He drew a right-angled triangle on which the two sides adjacent to the right angle were labelled with 3 and 4, and the side opposite the triangle was labelled with an *x*.

'Now,' he said, 'I want you to find *x*.'

'That's easy,' said Max, beating Rupert in raising his hand. 'It's five. You just use Pythagoras's rule. Three squared is nine. Four squared is sixteen. Add them together to get twenty-five, then taking the square root, you get five.'

'Good try,' he said, 'but not the answer I am looking for . . .'

While Chris spoke, Zamya had got slowly up from her seat, walked down to the board and pointed at the *x* written on the board. 'Here it is!' she exclaimed (figure 28).

Everyone laughed.

Find x

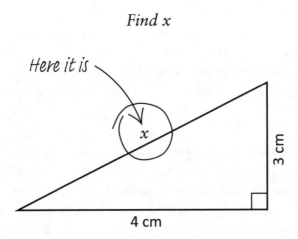

Here it is

x

3 cm

4 cm

Figure 28: Here it is!

Chris said, 'Exactly, Zamya. That is what I am looking for. Here it is! While the answer five is correct, the answer you gave is more important. Not just because it is funny, but because it steps outside the system. The answer emerges unexpectedly from the way I posed the problem.'

Chris told us that 'Here it is!' is a deeper, more profound reply to the question. It tells us something surprising, something we haven't thought about before. It breaks the unspoken agreement between teacher and student about how we use symbols like *x*, 3 and 4. The agreement to look only at the maths problem is artificial. It applies only within the narrow framework of the maths class.

'As Zamya will know,' he told us, 'this joke echoes the words of the Austrian philosopher Ludwig Wittgenstein.'

Chris told us that around the same time, at the turn of the century, that Fisher and Lotka had been struggling with questions about how to correctly measure and describe the phenomena around us, Wittgenstein had been trying to find an answer to a similar question: what can we say for certain about the world? An important insight into this problem can be found in the penultimate paragraph of his groundbreaking work of 1918, the *Tractatus Logico-Philosophicus*:

> *My propositions serve as elucidations in the following way: anyone who*
> *understands me eventually recognizes them as nonsensical, when he has*
> *used them – as steps to climb beyond them. (He must, so to speak, throw*
> *away the ladder after he has climbed up it.)*
>
> *He must transcend these propositions, and then he will see the world*
> *aright.*

Take, for example, the way Antônio reacted when we looked at complex cellular automata. Chris said he hoped that, in the future, Antônio will produce his own mathematical models of rainforest dynamics, which will explain how properties of ecosystems emerge from species interactions. These models will be a ladder for him to climb from the bottom up and view the system from above.

It is in this way that cellular automata and other models like this use Kolmogorov's definition of complexity. We strive to find the explanation that best reduces complexity. Once found, we have identified a new understanding of a phenomenon.

It is up to all of us to find the 'Here it is!' jokes that life plays on us, Chris said. We should step outside the system and realize a greater truth. Find the ways in which complexity emerges.

'But,' Chris now warned, 'Wittgenstein's ladder and the "Here it is!" joke provide an additional insight.'

They are about how we react after we have understood how a collective pattern emerges from individual interactions. Once we have seen how a pattern emerges, we have transcended our understanding of the phenomena we are studying. We see the joke that the world was playing on us. Now the complex has become simple.

This does not, however, mean that complexity has disappeared. Its form has just shifted with our new viewpoint. This is why, as Wittgenstein says, once we have climbed a ladder, we must be willing to throw it away. It is not the model itself that is important, it is the insight we gain from it. We should, once we have found some new understanding, now look for further ways to climb through complexity. The nature of complex systems, with their many sides and deep intricacies, means that they always throw up new problems and new questions. We can't ever satisfy ourselves with one view or one insight. We need, time after time, to restart the process, laugh at our previous attempts and start again.

Only once we realize the sheer scale of complexity can we start to navigate a way through its deepest secrets.

It's complicated

Throughout this book we have created models: models of our health and happiness; models of how couples argue or how we motivate our friends; models of chaos produced by our choices in life; and models of the social structures we see in groups, crowds and society. Through these models, Charlie and Aisha have found a way to argue less; Richard has developed a healthier relationship to sugary treats; and Jennifer has started to feel less isolated.

But . . . these models are never the end of the story. Each of these solutions has led to new challenges. Maybe the spark between Charlie and Aisha is ever so slightly diminished now that the arguments have gone? Maybe Richard has become sanctimonious as he points out how others can't control their urges? Maybe Jennifer's study of social structures made her feel a new sense of detachment, where she sees her fellow human beings as simply interchangeable points of larger, indiscriminate wholes?

What Ubuntu, 'Here it is!' and Wittgenstein's ladder tell us is that dealing with complexity is not just about finding the right model. It is also about a willingness to step out of that model and see the way it has changed us. We need to laugh at the 'Here it is!' joke we have played on ourselves, to throw away the ladder that took us upwards and see that we have changed and that we have changed other people around us.

The search for complexity is never-ending and stretches in all directions. We have already seen how it has taken us upwards, to look at how we interact as part of systems which are larger than ourselves. Now, in the final chapters of this book, we will turn around and travel in the opposite direction. We are going to finish our journey by looking inwards, to look at the complexity within ourselves.

Almost always complicated

To begin this final part of our journey, we return now to the International Congress of Mathematics in Nice in 1970.

Over lunch, several of the younger members of the Bourbaki group, the French purist mathematicians, made sure they sat down at Kolmogorov's table. They weren't convinced by his presentation and wanted him to start his explanation again, from the very start.

Kolmogorov was happy to oblige. He wrote the binary string, a sequence of bits

0000000000000

on his paper napkin. He explained that this string can be represented by writing

'Write zero 13 times.'

The statement isn't in itself shorter than the original string, but it is easy to use it to create very long strings. For example, while writing out 1,378 zeros would take a lot of space on a napkin, the phrase

'Write zero 1,378 times'

could be written down very succinctly. The same is true for periodic strings. For example

101101101101...

could be represented as

'Write 101 a certain number of times.'

All we need to do is find the part that repeats, 101 in this case, and then write down the number of times it repeats. This, Kolmogorov said, was the essence of his measure of complexity: a repeating binary string, or any string with a pattern that can be described using an algorithm has a low value of complexity which can be referred to as K.

Kolmogorov now told them to imagine the most complex string of 1s and 0s possible. Under his definition, this was a string that could only be described by describing the string itself. As an example, he wrote

0100100101110

on the napkin. He explained that the string above has thirteen bits, so its length is thirteen. Its K is also thirteen, because there is no way of shortening further than the number itself. The question he asked himself, he told them now, was how often such strings occurred. How many of all the different strings we can write down cannot be given a simpler explanation than simply writing down the string itself?

'Such strings must be very rare?' speculated one of the Bourbakis. 'We should be able to find patterns in most strings, shouldn't we?'

Kolmogorov's smile broadened. 'Well, if you knew me, young Bourbaki, you would know that what has defined my style as a mathematician is not so much answering questions but instead knowing the correct questions to pose. And your question turns out to be the wrong one. I hypothesized instead the opposite: in the case that long strings are almost always complicated, in the sense that there is no way of shortening them.'

To illustrate his point, Kolmogorov started writing down a long string of 1s and 0s on the napkin.

1010101100010001110010000011110010111011110000101

He had, he said, come to believe that strings like this one, something written down off the top of his head, can't be simplified or shortened. His hypothesis was, he told them, that once the strings get long enough, the probability that you will find a pattern becomes zero.

For Kolmogorov, this hypothesis felt true, but it was left to a young researcher from Sweden, Per Martin-Löf, who had visited Kolmogorov's research lab between 1964 and 1965, to prove it. Martin-Löf developed a test for complexity which involved systematically trying to find patterns within a binary string within different-sized chunks. He showed that strings which passed his test, which had no discernible pattern no matter how a string was broken down into chunks, were complex in the way defined by Kolmogorov: there was no way of shortening the description of them. Moreover, Martin-Löf found that for sufficiently large string lengths (above, say, about a few million bits or so) there are many more binary strings of a length that cannot be shortened than those that can be shortened. In fact, almost all sufficiently long binary strings are impossible to shorten.

'This means,' Kolmogorov said, 'that complexity is the rule, rather than the exception. Once the strings of 1s and 0s are long enough, they are nearly all impossible to shorten.'

The patterns that we cannot simplify far outnumber those for which we can provide a simple explanation.

Who am I?

The big question 'Who am I?' is one that most people ask themselves at some point in our lives. Some of us go on to spend a whole lifetime trying to find the answer, while others spend just as much time trying to avoid even considering the question. Whatever we do, though, the question is always there.

One starting point for finding an answer, the one which we take now, is to think of ourselves as a string of numbers and then ask, using Kolmogorov's approach, how complex that string might be. Is there a way to capture the essence of you in a single phrase or expression?

Inside your brain there are 86 billion neurons. That might sound like a big number but, in fact, it isn't. Not if you think about the number of ways your brain can be configured. Neuroscientists refer to neurons firing; a shorthand for describing them sending electrical signals to each other. Any one neuron can be in one of two states, either on or off, firing or dormant. Two neurons can be in four different configurations: both of them off; the first one off and the second one off; the first one on and the second one off; or both on. In terms of binary strings. these four alternatives are

00, 01, 10 and 11

Add a third neuron and you have eight possible configurations.

000, 001, 010, 011, 100, 101, 110 and 111

A fourth neuron gives sixteen possible configurations. And so on. Each time we add a neuron we multiply the number of possible different configurations of our brain by two. Thirty-two neurons have 2 to the power of 32, which is 4.3 billion, possible configurations.

To find all the possible neuronal configurations inside our skulls we would need to multiply 2 by itself 86 billion times. That is a quite big number – equal to 10 to the power of 25.9 billion. To write the number corresponding to the state of your brain would require a string of 25.9 billion digits.

And even this is a vast underestimation of its complexity. Although it is the neurons that do the firing, it is the hundreds of trillions of synapses that carry messages between the neurons. So, a proper estimate of brain complexity would be, at the very least, 10 to the power of 100 trillion. Just try to think how that number would look: a 1 with a trillion zeros written after it. It is, quite simply, impossible to imagine how many different ways your brain can be configured.

There is a myth that we only use 10 per cent of our brains; it originated from a quote incorrectly attributed to Albert Einstein. It was later peddled by spoon-bending 'psychic', Yuri Geller, who implied that the unused 90 per cent might allow us to deform small metal objects. Geller's ideas were, of course, total nonsense. But it is true that our brain experiences only a tiny fraction of its possible configurations. Imagine, for example, that your brain enters a new configuration every one millionth of a second. Then, assuming you live to be one hundred, your brain will experience less than 10 to the power of 16 different configurations (that is, 100 years × 365 days × 24 hours × 60 minutes × 60 seconds × 1 million configurations: $100 \times 365 \times 24 \times 60 \times 60 \times 1000000 = 3,153,600,000,000,000$, which is 3.155×10^{15}, which is less than 10^{16}).

Again, 10 to the power of 16 might sound like a lot, but if we divide it by 10 to the power of 25.9 billion, to get the proportion of configurations your brain has actually used, then the result is mind-bogglingly tiny. In fact,

$$\frac{10^{16}}{10^{25,900,000,000}} = \frac{1}{10^{25,900,000,000-16}} = \frac{1}{10^{25,899,999,984}}$$

So, over the course of a long life, using all of the 10 to the power of 16 different configurations, your experience does not even come close to encompassing all the possible states of your brain.

Charlie has been thinking a lot recently about who he is as a person. In contrast to Jennifer, who looked outwards to her social relationships with other students, Charlie has turned inwards to try and understand who he is. He started by thinking about what it is that defines him. He thought about what would happen if he wrote down all the words that have ever been said to him, or that he has heard in films, on the radio or on YouTube. He takes an educated guess at about 12,000 words per day, which is probably an underestimate – but he just wants to get a rough idea. Each word consists of an average of five letters, which can be encoded in eight bits (using ASCII code for example). This means that Charlie has experienced $34 \times 365 \times 12,000 \times 5 \times 8 = 5,956,800,000$ bits of information.

Since each of these bits can take a value of either 1 or 0, this means that there are roughly 2 to the power of 6 billion different possible strings of information that a thirty-four-year-old like Charlie might have experienced. In decimal, this is roughly 10 to the power of 2 billion. For context, the number of particles in the universe is usually cited as being 10 to the power of 80, which is a tiny fraction of 10 to the power of 2 billion. And this is just information he has received in the form of spoken words. Add to this what Charlie reads, what he sees, the sounds he hears, the smells and the tastes he has experienced, and his potential life experiences multiply even more.

Regardless of whether we perceive ourselves as either the make-up of our brains or the extent of our experiences, we would need a very long string of numbers to summarize that description. Charlie is a string consisting of billions upon billions of digits. He has experienced one out of an almost unimaginable variety of ways in which his life could have been. Mathematicians call the length of the string its dimension. Each and every one of us is a many-billion-dimensional object.

Kolmogorov's complexity theory, as described in the previous chapter, tells us that the vast majority of high-dimensional strings do not have a simpler explanation than one which involves writing out the string itself. This result is theoretical, but it is worth bearing in mind when we look at ourselves as a string of numbers. The strings encoding the blinking neurons in your brain and the vast ocean of your experiences are billions of digits long. Is there a short computer program that can reproduce who we are as individuals? Is there a low-dimensional representation of a person?

In an attempt to know who he really is, to somehow find his essence, Charlie has spent a lot of time searching online. Initially, he got into astrology. His star sign is Cancer, which, he read, explained why he felt shy and introverted in many situations. But then he knows this guy at work, who had the same birthday as him but is pretty much his total opposite: always laughing and joking, the life of the party. After this, and quite a few other similar experiences, he realized astrology didn't really make sense.

It was then that Charlie discovered online personality tests – like the Myers–Briggs Type Indicator, the DISC Personality Test and the Big Five Personality Test – which claimed to allow him to find out what type of person he is. These tests consist of about thirty or forty questions about the subject. For example, 'Does seeing other people cry make you cry?'; 'At social events, do you start conversations?'; 'Do you finish projects you start?'; 'Are you sentimental?'; 'You believe pondering philosophical questions is a waste of time?' and 'Are you inclined to follow your head rather than your heart?'

After the test, an illustration pops up telling Charlie about his personality. The most recent test he completed had five categories – 'Extroversion vs Introversion', 'Intuition vs Observation', 'Thinking vs Feeling', 'Judging vs Prospecting' and 'Assertiveness vs Turbulence' – and the test provided scores between one and ten for each category. Charlie scored two in terms of extroversion, eight in terms of intuition, four in terms of thinking, three in terms of judging and six in terms of assertiveness. The site further classified him as an introvert: a person with a vibrant inner life, with profound emotional responses to culture but shy among others.

Tests which assign five personality traits measure five dimensions of a person. This allows 10 to the power of 5, or 100,000, different possible scores, which ensures that not many people will get exactly the same score. But five dimensions is a minuscule number in comparison to the 10 to the power of 1 billion potential variations of Charlie.

It is the vast difference in the dimension of a personality test (five) and the dimensions of our human experience (several billion) which means that Charlie's search for his essence through online tests is doomed to failure. The number of dimensions of our minds and experiences far outnumber the number of dimensions we can measure about ourselves. Kolmogorov's complexity suggests that it is very unlikely that a single dimension, or even five dimensions, can capture any particular person. There is no way to simplify the string that is Charlie.

Completing online personality tests can be a useful exercise. But to get the most out of it, we should first understand our own irreducible complexity. One trick that helps when doing these quizzes is to think about subtleties in the answers. For example, when Charlie scored two in terms of extroversion, he thought about the questions 'I initiate conversations' or 'I don't talk a lot' in the context of parties and organized social gatherings, which he doesn't enjoy.

Instead of placing himself at a party when he answers these questions, Charlie should try placing himself in other contexts. Is he also shy at work? How is he around his family? How does he feel about making presentations? Does this change when he does them over Zoom? How does he behave with his friends? Maybe he feels freer to express himself there? What about if he meets people over coffee, rather than drinks, where others become loud and dominate the conversation? Does he feel more comfortable talking about certain subjects such as football, about which he knows a lot? Maybe he is shyer when it comes to chitchat but finds it easier to open up when he talks about his work or his family?

Introversion itself has many dimensions. It changes in different circumstances. In order to understand our individual complexity, we need to open up the dimensions within any question we pose about ourselves. Instead of naively accepting the ways a personality test can narrow us down, we should instead strive to see how the questions themselves can open us up, how they allow us to see ourselves in different ways in different situations.

Charlie will never be simplified to one or a small number of measurements, no matter what an online test says or however much others try to simplify or categorize him. Likewise, it is dangerous to put other people in boxes; to see them as shy or confident, angry or bitter, clever or stupid, organized or muddled . . . All these dimensions of a person change with context and the situations they find themselves in.

The answer to the 'Who am I?' question is that there is no single you. You are billion-dimensional.

A life in short scenes

The evening after his presentation to the World Congress of Mathematics in 1970, Kolmogorov sat in his room and thought about who he was.

His early life had gone so fast. He thought back to how his talents had been 'discovered', as a nineteen-year-old at Moscow University. During his first-year lectures, his teacher, the influential Professor Luzin, made a particular assertion as part of one of the mathematical proofs he was demonstrating, telling the students they could check the assertion's truth as an exercise. Kolmogorov saw immediately that the assertion Luzin made was in fact false and wrote down a counter-example. Luzin was astounded, and asked one of his brightest PhD students, Pavel Uryson, to check that Kolmogorov was correct, which he was. Impressed, Uryson invited Kolmogorov to join his more advanced classes, where the younger student duly found further mistakes in the teaching, forcing Uryson to rewrite part of his lecture notes. This process of learning by questioning everything went on and, two years into his undergraduate studies, Kolmogorov was obtaining results which surprised not only his colleagues in Moscow but the entire international mathematical community.

Then, as now, Kolmogorov could not understand why others did not always see what he saw. It seemed to him that almost all his peers in Moscow understood mathematics better than he did. They would produce long, complicated proofs, sometimes hundreds of pages, which he would struggle to take in entirely. Once he got the ideas, though, he would often see that they could be vastly simplified. Paradoxically, when he explained his own, more concise thoughts to his teachers, they would grasp the gist of his idea quickly, often kicking themselves because they had missed the 'obvious' solution.

This pattern would come to inform Kolmogorov's opinion that

mathematics was a balance between the trivial and the impossible. He would spend weeks or months pondering how a result could ever be found, only to experience a sudden change of perspective that made it all seem so straightforward. This ability to change his point of view was the reason his peers viewed him as a unique talent, a genius even. Kolmogorov himself set little store in the praise they lavished on him. In school, he had preferred both biology and history over mathematics. A true genius, he thought, would have insight into the real world. As a PhD student in 1920s Moscow, he thought now, he had no such insight.

This lack of real-world experience was very likely why the return of Pavel Alexandrov, another of Kolmogorov's undergraduate tutors, to Moscow in 1929 had such a profound effect on him. Thinking back to that time, Kolmogorov couldn't quite understand how he (by now a final year PhD student) had found the nerve to invite the great Alexandrov (the first Soviet mathematician to travel abroad), whom he knew only formally, on a several-week-long boat trip down the Volga. But he had extended the invitation, and Alexandrov had accepted.

On the trip, Alexandrov told Kolmogorov about his extensive travels both in the Soviet Union and throughout Europe, with Uryson (the tutor who had first checked Kolmogorov's results). Alexandrov described how the pair worked on mathematics in the mornings, sitting side by side in whatever room they happened to have rented at the time. During their visit to Göttingen in 1923 they participated in afternoon seminars with the father of modern mathematics, David Hilbert, and spent the evenings in intense mathematical discussions with the great algebraist Emmy Noether and her cohort of students (affectionally known as the 'Noether boys').

What caught Kolmogorov's imagination most were not the stories of illustrious German mathematicians but when Alexandrov talked about how he and Uryson walked and swam together. They had hiked through Norway, bathing in every bay and fjord along the way, no matter how cold the water was. They spent whole days lying in the sun, discussing not just mathematics, but literature and music. Pushkin. Dostoyevsky. Goethe. Beethoven. Tchaikovsky.

The pair also visited the cities of Europe. Alexandrov spoke about one night in particular, in late July 1924, when he and Uryson had been staying in a budget hotel opposite the Sorbonne in Paris. Coming back to their room after dinner, they stepped out on to their small balcony. Alexandrov described how the whole of Paris stretched out before them in a fading evening glow and, from a garret window opposite, they heard someone playing a Beethoven piano sonata. It was that memory, a moment of true love, that Alexandrov would, he told Kolmogorov, forever hold in his memory.

It was this type of life that Kolmogorov yearned for. A life of experience. Of mathematical mornings. Of swimming feely in lakes. Of poetry and music. Of travel. Of friendship. Of love . . .

The effect on Kolmogorov of Alexandrov's stories was made all the stronger because he knew that the evening in Paris was followed only a few weeks later by the most terrible, almost unspeakable, event.

It was 17 August 1924. Uryson and Alexandrov had now travelled to Brittany and were as usual spending the morning working in the hut they had rented in Batz, a small fishing village right on the coast. Absorbed in their work, they continued to do mathematics without a break until 5 p.m., much later than usual, before going to the beach for a swim. As they waded into the water an uneasiness rose between the two mathematicians. Was it really safe to enter the water? Waves were crashing in on the rocks and they hadn't eaten since breakfast.

After initial hesitation, they looked at each other, took a deep breath and plunged into a smaller wave, swimming, under the surface, out to sea. When they surfaced, Alexandrov found himself further from the shore than he expected. And then, suddenly, he was carried away by a huge force. Another, much larger wave picked him up and took him tumbling all the way back to the shore. When he recovered himself and looked up, he saw Uryson floating, face down, in a half-sitting position, about fifty metres away. Alexandrov dived back in and pulled his friend to the beach. He tried to resuscitate him. But it was too late. In only a few minutes, Uryson was gone. His far-too-short story had come to an end.

Kolmogorov realized that the boat trip down the Volga was, for Alexandrov, in part, a reliving of his adventures with Uryson five years previously. And he was honoured that Pusya – as he would come to call Alexandrov as they became closer – started to confide in him, and was later thrilled when he suggested that the two of them should visit Göttingen and Europe together.

What an experience those travels had been. Like it was yesterday, Kolmogorov still remembers the thrill of Hilbert's praise when he solved the problem the German mathematician had posed about axiomizing probability theory, or the insights gleaned from evenings with Emmy Noether and at lunches with other famous mathematicians. The two travelled through the Bavarian Alps, visited Freiburg and swam in Lake Annecy in the French Alps before making their way through Marseille to the coast at Sanary-sur-Mer. But as wonderful as those moments had been, the best, Kolmogorov thought now, were those that were still to come.

In 1935 they bought a summerhouse, a dacha, in the small village of Komarovka. Kolmogorov thought back to their blissful routine. Three days were spent in Moscow, four in Komarovka, one of which was devoted entirely to physical recreation – they would ski, row, or take long, long hikes. On sunny March days, Kolmogorov reminisced, they went out on skis wearing nothing but shorts, for as much as four hours at a stretch. They loved swimming in the river just as it began to melt. Andrej swam only short distances in icy water, but Pusya always swam much further. When skiing, it was the other way around. Kolmogorov could ski, in just shorts, over incredible distances. In the evenings they would listen to music together, often with students and colleagues who visited.

It was the mornings, after exercise, that were truly special, because then it was work. And for Kolmogorov, those years were a time of unlimited productivity. Part of this was a continuation of his work in pure mathematics, but more and more his efforts focused on how mathematics could be applied to the everyday world. It was ironic that these good times began during Stalin's purges and in the midst of the Second World War. But, like many of his peers in the West, Kolmogorov revelled in the challenges of war, applying his mind to intricate problems of modern combat, such as the effectiveness of artillery barrages.

In the 1930s, the Soviet Union lagged far behind Western Europe and the United States in their understanding of applied mathematics and statistics. Kolmogorov set about changing this. He devoured statistical theory, including the work of Fisher, linking his own abstract understanding of probability to the Cambridge statistician's practical view of maximum likelihood. He studied Lotka's paper on predator–prey models, and generalized it to model any number of different interacting species in an ecosystem. He started his own research programme on turbulence: the flow of streams, of waves on the surface of water, of the wakes left behind by a boat as it passes through surf and the disturbance in the air as an aeroplane takes off. When a boat moves very slowly, the water simply flows on either side of it, a stable pattern. When the flow increases, periodic waves are created in its wake. Kolmogorov's contribution was to demonstrate that at very high flows – those experienced at the tip of an aeroplane in flight, for example – the eddies will have random sizes, creating unpredictable turbulence. In other words, he showed how increasing the speed of the boat led to a transition from stability, through periodicity to chaos (the idea a young Stephen Wolfram would pick up decades later when defining his four classes).

By the early 1960s many USSR academics believed they were now ahead of their USA counterparts in their understanding of how mathematics could be used to model the physical world. The space programme was part of the evidence for this, but they also cited how Kolmogorov's insights into turbulence had come fifteen years before Edward Lorenz's application of chaos theory to the weather. It was Kolmogorov, his younger colleagues claimed, who made the first connection between a deterministic process (the forward movement of an object through a fluid) and a random out-come (the turbulence generated in the object's wake). Assertions of precedent didn't interest Kolmogorov. He judged scientific articles purely in terms of the change of perspective they brought. The computer simulations Margaret Hamilton and Ellen Fetter had per-formed to create Lorenz's butterfly gave one such change in perspective.

And it was a change of perspective that brought Kolmogorov to his current view of complexity. He had read Shannon's work on information theory with great interest, but he was more interested in the meaning of patterns than he was in the unpredictability which entropy measured. Indeed, when Claude and Betty Shannon took turns to predict the text of the Jefferson biography, they had chosen passages at random from the book. They had viewed the six volumes as an endless source of words to be predicted, rather than a finite description of Jefferson's life. They had deliberately ignored the meaning of the work, never reading it from start to finish.

Kolmogorov's own life was, by the 1960s, full of meaning. That meaning, he saw now, had first arrived in the form of his friendship with Pusyas. Alexandrov had taught him how to appreciate every moment. How to use his hours wisely. Once Kolmogorov had understood that every action had a meaning, he was able to focus on what mattered most to him. The joy of teaching young and old, schoolchildren and PhD students. The pride of seeing a student suc-ceed in solving a difficult problem. The insight he found from hearing about the lives of others, the complexity of their emotions. The depth of the ideas he found when he read the poems of Push-kin or listened to the symphonies of Beethoven. The sensation of swimming in the lake and being suddenly struck by the sort of idea that can only arrive in the water. The feelings he had when he spoke of love with Pusya.

It was this meaning in his own life that produced Kolmogorov's insight: if he were to analyse Jefferson's biography, or Tolstoy's *War and Peace*, or Pushkin's poems, he would not sample random pas-sages, as the Shannons had done. He would look at the story, the structure of the prose, the melody of the stanzas. He would find the elements that gave these works their meaning.

Meaning, he realized, was to be found in the finite world of the here and now. It lay in how well we describe a situation. A person has a finite life and makes decisions about what they do with their time on Earth. Some people will lead boring lives, never travelling, never searching for deeper truths, while some will lead rich lives, always seeking adventure, always learning and interacting with others. Uryson's life, for example, was short in years but full in content. It was this way of living which Alexandrov had brought to him, filling his life with a rich diversity of feelings and emotions.

Now, sitting late at night in the hotel room in Nice, he looked over to the bed, where Pusya lay in a deep sleep, tired after a full day of mathematical thought. His story is long, taking many turns, involving many friendships and many intellectual challenges. The complexity of a person does not lie in the abstract or the infinite, Kolmogorov thought, but in the finite description of what they do with their life. The more complex a person is, the harder we have to work to describe them.

An indescribable explanation

It was our final evening. The last time we would all be together in Santa Fe.

We'd agreed to meet in a quieter, fashionable bar in downtown Santa Fe. A stark contrast to the loud nightclubs and sports bars we had visited in previous weekends.

When I arrived, Parker was there, talking to Zamya. The two of them sat at the same table as my friends – Esther, Madeleine, Antônio, Alex, Rupert and Max – but a small distance away from the others. They were deep in conversation.

'What is Parker telling her about?' I asked Esther, eager not to miss out. I still felt there might be something more to be learned, even after Chris's brilliant 'Here it is!' finale.

She looked at me. 'You really still think that Parker knows it all, don't you?' she said with a slightly condescending smile. 'Look a bit closer.'

When I looked, I saw that Parker wasn't telling Zamya anything. It was Zamya who was doing the talking. Parker was trying to interject, and getting more frustrated, but she just continued, calmly explaining her point.

'Probably just a lot of postmodernist bullshit,' interjected Rupert. 'No wonder he is getting annoyed with her. Philosophers going around saying that science is wrong, as usual. That maths is just as valid as horoscopes or religion. Thinking they can amaze us with their "Here it is!" all the time.'

Rupert started telling the rest of us about a recent stir created in philosophy by Alan Sokal, a physicist from New York. A year earlier, in 1996, Sokal had submitted an article to the academic journal *Social Text* with the title 'Transgressing the boundaries: towards a transformative hermeneutics of quantum gravity'. The article contained an unsubstantiated claim that 'physical reality . . . is at bottom a social and linguistic construct'. It went on to talk about how the validity of psychoanalysis had been confirmed by work in quantum field theory, that the axiom of equality in set theory has analogous concepts in feminism and that quantum gravity has profound political implications.

The journal published the article, but three weeks later Sokal revealed that it was, in fact, a hoax. His aim had been to use the publication to criticize a movement in philosophy known as postmodernism, which to Sokal (and to Rupert), was a mixture of overly complicated words and pretentiousness. Sokal wanted to show that scientific jargon could be used in the humanities to persuade readers that anything goes; even that there was no such thing as physical reality. Rupert laughed as he told us about the hoax, and how Sokal's article had revealed the postmodernists for what they were: pseudo-scientific charlatans.

'Compared to postmodernism, these last weeks have been much more down to earth,' Rupert concluded. 'In fact, I have to admit I have learned a thing or two during this summer school. But I'm sure what Zamya has to say is a step too far . . .'

Zamya, apparently having finished with a now somewhat despondent-looking Parker, turned towards the rest of us. Rupert's voice had projected throughout the bar and was difficult to ignore.

'Sokal offered only arrogance in his journal submission,' she said, turning to him. 'Sadly, he lacks seriousness as a thinker.' Zamya explained that denial of physical reality was one of many philosophical positions which could be investigated. The point of a postmodernism critique was to always question the assumptions we make, to be modest about what we can say about what we know.

Rupert scoffed, 'So you think that Sokal is more arrogant than philosophers like Wittgenstein, who avoid all responsibility by climbing out of the room with a ladder?'

The rest of us laughed, thinking back to what Chris had told us about the Austrian philosopher's words. Zamya smiled too but, after some thought, she responded carefully.

'It depends how you define "arrogant",' she said. 'Some people might consider you arrogant, Rupert. The rest of us have come here to learn. You have come here to neutralize the danger to your own comfortable theoretical world. You first perceived Parker's lectures as a threat; now Chris has become your hero. And, here you are, defending those two against the next perceived danger. You seem to never learn that there is always one more layer to complexity.'

Rupert was sheepish, not knowing quite what to say. And, when I looked round, I noticed that Parker had gone. Initially, I had thought he was just ordering another drink, but now I realized he had left directly after Zamya had turned to talk to us. It didn't look like he was coming back.

Esther took up the challenge. 'I am sure you are right in some sort of abstract-philosophy type of way, Zamya: that scientists are arrogant. But let's face it, we provide answers to questions. That is what I was explaining to David the other day when we were at the institute. Parker doesn't understand it all, and neither, bless him, does Rupert, but that doesn't mean that anything goes.'

'Addressing those points was exactly my reason for talking to Parker,' Zamya said to Esther. 'I told him what I thought was missing from the lessons here. He and Chris haven't been clear on the limits of their approach to complexity.'

Zamya explained that Sokal's article merely demonstrated that one set of reviewers at a particular scientific journal could be fooled by an article written in bad faith. The real problem was how scientists failed to acknowledge a more fundamental limit in their own work: that we can never know if a theory is truly correct. Likewise, we can never be sure we have found the simplest explanation for a complex phenomenon. Many scientists act as if they could uncover the true nature of reality, Zamya said, but they never will, they can never be sure . . .

'I don't see why not,' replied Esther, 'All I need to do is find the correct algorithm or computer code for calculating the complexity of any given situation. I might not know what that algorithm is right now, but eventually I'll find it and will be able to tell you the answer.'

'And that is where you have a problem . . .' Zamya answered. 'Complexity can't be computed! You will never find that algorithm, Esther. Let me explain . . .'

Zamya reiterated how there were – just as Chris had presented – many ways in which simple rules could explain the behaviour of complex systems. But she also stressed that Kolmogorov and Martin-Löf had shown there were also many more examples of complex strings that could never be reduced to simple rules. In an ideal world, we would, as Esther had suggested, try to find an algorithm or method for identifying whether a particular system was truly complex (and couldn't be further simplified) or was one of those which could be explained in a simpler way. What she meant by saying that complexity couldn't be computed was that the existence of such an algorithm or a machine – one designed to calculate the complexity of any given string – was a logical impossibility.

To prove this statement, Zamya said she would use a technique called proof by contradiction. When proving any statement by contradiction, the approach is to first assume that the statement is true and then show that this assumption leads to a conclusion which we know cannot be true.

In this case, Zamya said, let's assume that Esther has a machine that allows her to calculate the Kolmogorov complexity of any given string. Let's also assume that the computer code for Esther's machine is 1 million bits long. Now imagine that David asks Esther to find a string with, for example, 2 million bits which has Kolmogorov complexity of 2 million bits: it is as complex as its shortest description. To do this, Esther can use her machine to calculate the complexity of one 2-million-bit string after another until she finds one which has a complexity of exactly 2 million.

'How do you know that such a string even exists?' asked Rupert.

'It is true,' replied Zamya, 'that she might have to test a lot of strings to find it.' She reminded us that there are 2 to the power of n ways to create a binary string of length n. This is because each bit in the string can take one of two values, 0 or 1. For example, a string of length $n = 3$ can take values 000, 001, 010, 011, 100, 101, 110 or 111, which is $2^3 = 8$ ways. So, there are 2 to the power of 2 million such strings of length $n = 2,000,000$ for Esther's algorithm to check. But it is (assuming Esther really does have a Kolmogorov complexity calculating machine) possible in principle. And all Esther needs to do is write an additional piece of code that says 'for each string, apply my machine to it'. As Martin-Löf proved, there are many strings which are complicated of any given length. Eventually, Esther will find a 2-million-bit complex string with her machine and can give it to David.

The string David receives, Zamya explained, is the contradiction. On the one hand, Esther has told him that it has a complexity $K = 2,000,000$. But on the other hand, Esther has used her machine, that is 1 million bits long, to produce this supposed $K = 2,000,000$ string. By Kolmogorov's definition, the complexity of a string is the shortest algorithm for generating that string. In Esther's case, her algorithm is just over 1 million bits long (the length of her machine plus a few extra bits for the code to loop over all the strings). She is thus claiming that the Kolmogorov complexity of the string she has produced is $K \approx 1,000,000$. At this point, David knows that Esther must be mistaken: she can't possibly claim that $K = 2,000,000$ and $K \approx 1,000,000$. Her machine cannot exist!

This was quite a lot for me (and the others) to take in, but I thought I might have found a hole in Zamya's argument. 'What if Esther's machine is greater than 1 million bits long?' I asked.

'It doesn't matter,' Zamya replied, 'if the machine exists, then Esther should be able to tell you how big it is. All you, David, have to do in order to get a contradiction, is ask Esther how big her machine is. Then ask her for a complex string of a length of, say, twice that of the machine.' If Esther provides an answer, she is contradicting herself. She is claiming to have a string which has almost twice the complexity of the machine used to calculate it! By Kolmogorov's definition, this is not possible. Even if Esther doesn't know exactly how big her machine is, in order for it to exist it must fit into the memory of her computer – it must be finite. In which case, David can just keep asking for larger and larger examples until Esther is forced into contradiction.

'The way in which this contradiction arises is similar to how we use words like "indescribable" or "ineffable",' Zamya said. 'When we want to tell another person how we feel; how much we appreciate their help; or even how much we love them, we might say that our feelings cannot be put into words.'

Zamya explained that accurately describing another person is akin to finding the shortest possible way of explaining our feelings about them. In many cases, no matter how we try, we can't find a way to express ourselves, so we instead say that our feeling is simply indescribable. But this is itself a description of our feelings, a very concise one at that. This produces the same contradiction that Esther's Kolmogorov complexity calculating machine created: we describe a person by saying they are indescribable.

We sat in silence for a minute, taking in what Zamya had said.

It was Alex who broke the silence. 'I tell you what is in-effing sad,' he said. 'That we are sitting around in silence on our last day in Santa Fe. I know a nightclub! Let's go dancing . . . Let's go out on a bang . . .'

The least words are the deepest

Back in modern-day London, Aisha has been telling the other nine friends, all of whom have got together for dinner, about the video she made at work about homelessness.

After she has shown them the final product, Becky asks, 'Do you really think you can sum up people's lives in a short video or in a few words?'

Antony smiles and says, 'Well, if I was to sum you up in a few words, Becky, it would be "the girl asking the questions". And now is no exception.'

The others laugh, thinking about how Becky recently told them that the secret to being a good listener lies simply in the choice of questions.

Antony continues, 'For me, it is easy. I'm the source of all chaos and (hopefully) love in Nia's life.'

'That you definitely are,' smiles Nia. She tells the others that she has finally admitted these last weeks that she's a 'bit of a control freak, who is finally learning to let go'.

'You're similar to me, John,' Nia says. 'You are always there for us, helping us get back on track. Bouncing us like basketballs in what you think is the right direction. But, you know, you can let go now and again too . . .'

Suki confesses that she is still the consummate follower of trends, and the best trend she has followed recently is Sofie's new fitness regime. Richard agrees. He was slow to get into the healthy living that the others adopted – and he's still a 'secret chocolate-cake addict' – but he's determined to keep his consumption more stable in the future.

'Seriously, though . . .' Charlie says, 'I have thought about this question a lot recently. Are we really just such simple caricatures?'

'You certainly aren't a caricature,' says Aisha. 'I know we have had our ups and downs, but we have worked hard to find a better way of communicating, to cut down on the shouting. And what I have realized is that to me, as my husband, you are unique; uniquely mine. And I would never change you.'

A small 'aahhhh' spreads through the ten friends and they sit in silence for a minute. Charlie is thinking.

'Thank you, Aisha,' he says at last. 'Those words mean everything coming from a person who cares so much about everyone. To be uniquely yours is the greatest privilege anyone could have.'

'What about you, Jennifer?' asks Becky (her again). 'You're very quiet.'

Jennifer tells the others that her semester as a student has given her a chance to see the social world from a distance. When she was working in London, she felt trapped in her role as a 'daily commuter'. But now she's realized that our roles change depending upon the social system we're a part of. She can now see herself in a role as the 'eternal student' instead, the kind of person always looking for new knowledge. She is learning, she says, to understand every situation for what it is.

We can summarize the friends – Jennifer always learning; Nia as a control freak; Becky as an inquisitive listener; and so on – in a way that is much more accurate than the list of numbers (age, income, lattes drunk and gherkin preference) with which we started this book. These descriptions better capture the essence of a complex person. We should strive for a simple idea that allows a larger idea to emerge.

Ultimately, however, and this is the lesson Zamya taught us in Santa Fe, there is no way of knowing whether the way we have described a person or a social situation is the best. We know that any short description will omit details, but we don't know how important those details are. These descriptions change with context. Charlie is a different person at work, or at a party or when he is on his own thinking about his own existence. Even Aisha, who knows so many things about him, will never fully comprehend his complexity.

We can never fully understand another person. Perhaps we can't even fully understand ourselves.

Four Ways

It can feel overwhelming at times: listening to your own internal arguments, counterarguments, worries and self-criticism. Why does your boss treat you differently to the others? Why are you always bickering with your siblings? Why can't you keep up with the long list of things you want to do in your life? Why didn't you make better choices in the past? What will you do in the future? Why don't you feel as smart as your friends? Is one of them trying to make you feel stupid? Why is your colleague failing to do their job properly, and they don't seem to care? Why don't your teenage children listen to you? Why are your parents always moaning?

It is when these thoughts, and others like them, build up that you should pause to consider the ways in which you think. You should analyse which thought processes get it right and which of them mislead you. You should remember that while there are many problems in life, there are four ways of thinking.

There is thinking based on numbers: how often does it happen to you and to others? Do your research. Collect the evidence.

Thinking based on interactions: how do you respond to each other? Find a way of breaking the negative cycle.

Thinking based on chaos: is it better to let go or take control of a situation? If you let go, then embrace the randomness. If you take control, then prepare your strategy as if you are landing on the moon.

And there are thoughts about complexity. While we can use the first three ways of thinking to handle conflicts with others, remember that we are all part of a much larger social system: family, work and society. And we all have our own innermost feelings that we can often never fully understand. Try to see everyone as an individual, by finding the words that best capture who they are.

Shape your own thoughts in ways which take you closer to the truth. But also know that, because we are all indescribably complex, there will always remain things that are beyond our comprehension. Don't worry about this; there is nothing you can do about it. Instead, let the multitudes and mystery each and every one of us contains inspire you. Use the finite time you have here in this world to enjoy the knowledge that there is always something new to discover in others and within yourself.

A worthy life

For Kolmogorov, a week of study should always start with a walk. A long walk. Made together.

A group of around ten to twelve PhD students were picked up in Moscow and taken to his beloved dacha in Komarovka by a transport vehicle commandeered from the Soviet state. Sandwiches were provided, and the next day the group set off on a long walk through the surrounding countryside.

Kolmogorov would talk with each student in turn. If he saw a student drop away from the rest of the group, he would wait for them to catch up. Then he would walk with the student, asking questions and listening to the answers. Kolmogorov never started the discussions by asking about mathematics. He talked about their lives, their interest in sport, whether they played chess, asked about their taste in music, what they did in their free time and their relationships with others. He listened carefully to their answers.

Many of the students were relieved by the small talk, terrified as they were of his mathematical mastery. They feared that Kolmogorov might, at any time, almost as an aside, find an error in what they had believed was an airtight mathematical proof, or find a much simpler, more elegant solution to a problem they had answered in a long-winded fashion. The work of a whole PhD thesis, they fretted, could be rendered meaningless by a single comment. But equally, Kolmogorov could provide the key idea the student might need to finish their work. Now that he had reached old age, the great man would regularly fall asleep in seminars – but he still retained his magic: he would often suddenly and unexpectedly wake up and explain how a problem that a student was facing could be solved simply through a change of perspective. If they were lucky, he would stand up and sketch the solution on the board, so they could use it in their own assignment. On other occasions he would just mumble a few thoughts before leaving the room. Either way, the students would spend weeks trying to decipher the meaning of what the professor had communicated.

Kolmogorov seemed not to fully appreciate the extent of the power he possessed. He considered much of what he said about mathematics to be obvious so would instead focus on posing problems that lay just on the edge of what he took for granted. Possibly because formal mathematics came so naturally to him, Kolmogorov put value not in rigour but in personal intuition. He often claimed that every person was provided, at an early time in their life, with a unique perspective of the world. In his case, he believed this arrived when he was fourteen years old. That his unique perspective happened to be (in my words, not his) that of a mathematical genius slightly superior to the collective efforts of all twentieth-century French mathematics was for him a trivial detail. Rather than focusing on himself, he wanted to know about the unique complexity that lay within everyone he met. He wanted to know what was going on inside other people.

Kolmogorov saw little difference between talking about maths and talking about life. He wanted his students to use advanced mathematical concepts just as freely as they would use their common sense. He often said that his guiding principle was sincerity. 'Our mission,' he told the other professors at Moscow University, 'is to find sincerity and to cultivate it.'

In 1986, Kolmogorov is an old man and has been ill for some time. Alexsandrov had died four years earlier, and Kolmogorov will soon join him. But today he has focused all his energy on the walk. The shining sun seems to have given him new life. He walks with the students, chatting, listening and offering advice.

Towards the end of the walk, Kolmogorov suddenly sets off in the direction of the nearby lake, leaving the others behind.

One of his students follows him and, when they get to the banks of the lake, he asks, 'What is it, Andrej Nikolayevich?'

Kolmogorov looks up towards the sky and says, 'I am proud that I lived a worthy life.'

The student says nothing. The other members of the group catch up with them and stand behind the professor in silence, staring at the same empty point.

Kolmogorov knew, looking at the endless distance in the sky, that there would never be an answer to all the complexity he had talked about, thought about and lived out. But his friendship with Pusya, the way he engaged others in teaching, the companionship of his students, and the discussions of life and mathematics were all ways of moving towards clarity. The meaning of life, the one that he had found, involved enriching his own internal thoughts and engaging with the complex internal lives of others.

This was what made his life worthy. Words spoken sincerely to each other. Meeting and listening to those near to him. Looking upwards and forwards at the sky together. Hoping and believing that the truth will move just that tiny bit closer.

Acknowledgements

I want to start by thanking Lovisa. Our discussions and debates provide a lot of the material in this book and the love we have is daily support to everything I do. Thank you.

The majority of this book was written in a pandemic bubble. I want to thank Elise for walking (together with Tobias and Ruby) and talking with me every day and Henry for always asking questions, many of which I still can't answer.

The characters in the story from Santa Fe are fictional. I did go to the Santa Fe summer school in 1997, and I met many interesting people there. The characters in this book are a mixture of these summer-school participants and other researchers I worked with at the turn of the millennium. If you feel I am representing you in one of the Santa Fe characters, then I probably am!

One very special person in this context is my PhD supervisor, Dave Broomhead. He did, like David's supervisor, always encourage me to ask questions. He also practised this skill himself, listening carefully and helping others develop in their own way. I think about Dave often, and he is greatly missed.

Thank you, Mum, for your detailed reading of an earlier version of this book, to Dad for several valuable ideas and Colin for his feedback: 'Book is good, head hurts from reading it though.' I am hopeful that, Ruth, you might actually read this one when it comes out!

Thank you to Casiana Ionita and Edward Kirke for working so hard to allow me to write what I want to write. Your patience, detailed editing, ways of challenging me to do more and constant support are the reason this book exists. Thank you to Chris Wellbelove for taking me longer and longer on this journey into mathematical writing. And thanks to Sarah Day for careful copy-editing and subtle changes that have greatly improved the text.

Notes and References

These notes provide references and further material for each of the chapters. For more in-depth explanations of the mathematics used, see the webpage https://fourways.readthedocs.io/

Four Ways

Stephen Wolfram, *A New Kind of Science*, Wolfram Media, Inc., 2002

Class I: Statistical Thinking

Very average friends

Commuting times taken from Glenn Lyons and Kiron Chatterjee, 'A human perspective on the daily commute: costs, benefits and trade-offs', *Transport Reviews* 28, no. 2 (2008): 181–98

Daily time spent watching TV per individual in the United Kingdom (UK) from 2005 to 2020: https://www.statista.com/statistics/269870/daily-tv-viewing-time-in-the-uk/

Median time length of penetrative intercourse from a study of people living in the UK from Marcel D. Waldinger et al., 'Ejaculation disorders: a multinational population survey of intravaginal ejaculation latency time', *Journal of Sexual Medicine* 2, no. 4 (2005): 492–7

K. M. Wall, R. Stephenson and P. S. Sullivan, 'Frequency of sexual activity with most recent male partner among young, Internet-using men who have sex with men in the United States', *Journal of Homosexuality* 60, no. 10 (2013): 1520–38

Life expectancy at birth in UK: https://data.worldbank.org/indicator/SP.DYN.LE00.IN?locations=GB

Births per woman in the UK: https://data.worldbank.org/indicator/SP.DYN.TFRT.IN?locations=GB

Notes and References

Happiness data from John Helliwell et al., 'Happiness, benevolence, and trust during COVID-19 and beyond', World Happiness Report: 13

A likely answer

'. . . final part of the Mathematical Tripos': Part II at that time was the third part of the Tripos (called Part III in the modern Cambridge syllabus). R. A. Fisher's position as a Wrangler is recorded in the Historical Register of the University of Cambridge, Supplement, 1911–1920

'. . . by multiplying the likelihoods of all the answers together': To see the logic of multiplying, imagine I am throwing a dice and I want to know the probability that I get a six on the first throw but not on the second throw. Well, the probability of getting a six the first time is 1 in 6, and the probability of not getting a six on the second is 5 in 6, so the probability of getting the six and then not a six is $1/6 \times 5/6$

'. . . maximum likelihood estimate we still use in statistics today': For more details see John Aldrich, 'R. A. Fisher and the making of maximum likelihood 1912–1922', *Statistical Science* 12, no. 3 (1997): 162–76

In the online tutorial I show how to prove that no other value is more likely than 40 per cent. See https://fourways.readthedocs.io/ and click on 'A likely answer'.

Twelve extra years

David L. Katz and Suzanne Meller, 'Can we say what diet is best for health?', *Annual Review of Public Health* 35 (2014): 83–103

Martin Loef and Harald Walach, 'The combined effects of healthy lifestyle behaviors on all-cause mortality: a systematic review and meta-analysis', *Preventive Medicine* 55, no. 3 (2012): 163–70

Elisabeth G. Kvaavik et al., 'Influence of individual and combined health behaviors on total and cause-specific mortality in men and women: the United Kingdom health and lifestyle survey', *Archives of Internal Medicine* 170, no. 8 (2010): 711–18, p. 711

How do you take your tea?

The text in this chapter is based on Chapter 2 of the biography of Fisher by his daughter: Joan Fisher Box, *R. A. Fisher: The Life of a Scientist*, John Wiley and Sons, 1980

'At meetings he rails . . .': quotes from Ronald Aylmer Fisher, 'Some hopes of a eugenist', *Eugenics Review* 5, no. 4 (1914): 309

'no one knew what to do with a woman worker . . .': Edward John Russell, *A History of Agricultural Science in Great Britain*, Allen and Unwin, 1966

'Together with another colleague, William Roach . . .': The dialogue here is a fictional adaptation of an account by Fisher's daughter, in Chapter 5 of Joan Fisher Box, *R. A. Fisher: The Life of a Scientist*, John Wiley and Sons, 1980. The design is described on page 13 of Fisher's *The Design of Experiments* (Oliver and Boyd, 1935). I have written that Roach suggested the alternative set-up of a paired comparison for dramatic effect. In fact, it was just one of many alternative set-ups that were used for experiments at that time.

For more information see:

R. A. Fisher, 'The arrangement of field experiments', *Journal of the Ministry of Agriculture* 33 (1926): 503–15

Bradley Efron, 'R. A. Fisher in the 21st century', *Statistical Science* (1998): 95–114

A happy world

See https://fourways.readthedocs.io/ and click on 'A happy world'. For more details see https://worldhappiness.report

The happy individual

See https://fourways.readthedocs.io/ and click on 'The happy individual' for more details of the analysis described in this chapter.

The *USA Today* article referenced in this chapter is https://eu.usatoday.com/story/money/personalfinance/2017/07/24/yes-you-can-buy-happiness-if-you-spend-save-time/506092001/

The probability that we get exactly k heads in forty coin tosses is

$$\binom{40}{k}\left(\frac{1}{2}\right)^{40}$$

The probability that we get twenty-six or more heads is

$$\sum_{k=26}^{40}\binom{40}{k}\left(\frac{1}{2}\right)^{40}$$

which is equal to approximately 0.0403. Where the threshold probability for statistical significance should be set is (to say the least) somewhat controversial. But following a standard of 0.05 for a one-tailed test (of whether time-saving leads to more happiness), this would be considered significant.

Angry old man

'Fisher took umbrage . . .': Ronald Aylmer Fisher, 'Design of experiments', *British Medical Journal* 1, no. 3923 (1936): 554

'One colleague described him . . .': Leonard J. Savage, 'On rereading R. A. Fisher', *Annals of Statistics* (1976): 441–500

'One of Fisher's friends described him as . . .': H. J. Eysenck, 'Were we really wrong?', *American Journal of Epidemiology* 133, no. 5 (1991): 429–33

'Joan Fisher Box, witnessed first hand . . .': Quotes on pp. 392–4 from Joan Fisher Box, *R. A. Fisher: The Life of a Scientist'*, Wiley and Sons, 1980

'He thought social classes and nations . . .': Ronald Aylmer Fisher, 'Some hopes of a eugenist', *Eugenics Review* 5, no. 4 (1914): 309

'. . . soon found himself arguing against his own theoretical results': Ronald A. Fisher, 'The elimination of mental defect', *Eugenics Review* 16, no. 2 (1924): 114

'Fisher's Cambridge colleagues . . .': Reginald Crundall Punnett, 'Eliminating feeblemindedness: ten per cent of American population probably carriers of mental defect – if only those who are actually feebleminded are dealt with, it will require more than 8,000 years to eliminate the defect – new method of procedure needed', *Journal of Heredity* 8, no. 10 (1917): 464–5

'He dug out previously discarded data . . .': Ronald Aylmer Fisher, *Smoking: The Cancer Controversy: Some Attempts to Assess the Evidence*, Oliver and Boyd, 1959

For a review of what we now know about cancer and smoking see US Department of Health and Human Services, 'The health consequences of smoking – 50 years of progress: a report of the Surgeon General' (2014).

The forest and the tree

See https://fourways.readthedocs.io/ and click on 'The forest and the tree' for more details of the analysis described in this chapter.

'The talk is based on a study that Duckworth . . .': Angela L. Duckworth et al., 'Grit: perseverance and passion for long-term goals', *Journal of Personality and Social Psychology* 92, no. 6 (2007): 1087

'When grit has been tested in meta-studies . . .': Marcus Credé, Michael C. Tynan and Peter D. Harms, 'Much ado about grit: a meta-analytic synthesis of the grit literature', *Journal of Personality and Social Psychology* 113, no. 3 (2017): 492

'Experimental observations have shown that the growth mindset approach . . .': David I. Miller, 'When do growth mindset interventions work?', *Trends in Cognitive Sciences* 23, no. 11 (2019): 910–12

'. . . only around 1 per cent of the variance between people': Carmela A. White, Bob Uttl and Mark D. Holder, 'Meta-analyses of positive psychology interventions: the effects are much smaller than previously reported', *PloS One* 14, no. 5 (2019): e0216588

'. . . emotional intelligence explains only 3 or 4 per cent . . .': Carolyn MacCann et al., 'Emotional intelligence predicts academic performance: a meta-analysis', *Psychological Bulletin* 146, no. 2 (2020): 150

Class II: Interactive thinking

The cycle of life

Herbert Spencer, *First Principles of a New System of Philosophy*, D. Appleton and Company, 1876. Quote from p. 434, section 173, Ch. 22

The version of the chemical reaction described here is from the 1920 paper: Alfred J. Lotka, 'Undamped oscillations derived from the law of mass action', *Journal of the American Chemical Society* 42, no. 8 (1920): 1595–9

But similar ideas are to be found in Lotka's 1910 paper:

Alfred Lotka, 'Zur theorie der periodischen reaktionen', *Zeitschrift für physikal-ische Chemie* 72, no. 1 (1910), 508–11

Rabbits and foxes

See https://fourways.readthedocs.io/ section 'Rabbits and foxes' for an in-depth mathematical investigation of this model.

The social epidemic

See https://fourways.readthedocs.io/ section 'Rabbits and foxes' for an in-depth mathematical investigation of this model.

Instead of focusing on changing minds . . . : for a more complete discussion of best practice around debunking misinformation see Sander Van Der Linden, 'Misinformation: susceptibility, spread, and interventions to immunize the public', *Nature Medicine* 28, no. 3 (2022): 460–67

The following references were used for the examples in this chapter:

Frank Schweitzer and Robert Mach, 'The epidemics of donations: logistic growth and power-laws', *PLoS One* 3, no. 1 (2008): e1458

Sarah Seewoester Cain, 'When laughter fades: individual participation during open-mic comedy performances', PhD dissertation, Rice University, 2018

Richard P. Mann et al., 'The dynamics of audience applause', *Journal of the Royal Society Interface* 10, no. 85 (2013): 20130466

Harold Herzog, 'Forty-two thousand and one Dalmatians: fads, social contagion, and dog breed popularity', *Society and Animals* 14, no. 4 (2006): 383–97

Nicholas A. Christakis, and James H. Fowler, 'Social contagion theory: examining dynamic social networks and human behavior', *Statistics in Medicine* 32, no. 4 (2013): 556–77

Yvonne Aberg, 'The contagiousness of divorce', *The Oxford Handbook of Analytical Sociology* (2009): 342–64

A third law

Ronald Ross, 'An application of the theory of probabilities to the study of a priori pathometry: Part I', *Proceedings of the Royal Society of London. Series A, Containing papers of a mathematical and physical character* 92, no. 638 (1916): 204–30

Ronald Ross and Hilda P. Hudson, 'An application of the theory of probabilities to the study of a priori pathometry: Part II', *Proceedings of the Royal Society of London. Series A, Containing papers of a mathematical and physical character* 93, no. 650 (1917): 212–25

'This is now made to appear probable . . .': Alfred J. Lotka, 'Contribution to the energetics of evolution', *Proceedings of the National Academy of Sciences of the United States of America* 8, no. 6 (1922): 147

Alfred J. Lotka, *Elements of Physical Biology*, Williams and Wilkins, 1925

Cellular automata

See https://fourways.readthedocs.io/ section 'Cellular automata' to run the models described in this chapter.

The art of a good argument

See https://fourways.readthedocs.io/ section 'The art of a good argument' to run the model described in this chapter.

For more on integrative behavioural couple therapy see Andrew Christensen and Brian D. Doss, 'Integrative behavioral couple therapy', *Current Opinion in Psychology* 13 (2017): 111–14.

Class III: Chaotic Thinking

Always knowing the next step

This chapter is based on 'Oral History of Margaret Hamilton', interviewed by David C. Brock on 13 April 2017 in Boston, MA. See https://www.youtube.com/watch?v=6bVRytYSTEk

'Lorenz told her what he knew, handed her the instruction manual . . .': 37:01 minutes

'For them "girls were just to go out with" ': 47:00 minutes

Notes and References

El Farol

See https://fourways.readthedocs.io/ section 'El Farol' for a more in-depth analysis of the model described by Alex.

The original El Farol bar problem was proposed by Brian Arthur in 1994 in W. Brian Arthur, 'Inductive reasoning and bounded rationality', *American Economic Review* 84, no. 2 (1994): 406–11

The mistake

'. . . modifications, this hacking, as she called it, would fix her original mistake': Margaret H. Hamilton, 'What the errors tell us', *IEEE Software* 35, no. 5 (2018): 32–7

'When Lorenz got in the next day . . .': 38:21 minutes. 'Oral history of Margaret Hamilton', interviewed by David C. Brock on 13 April 2017 in Boston, MA. See: https://www.youtube.com/watch?v=6bVRytYSTEk

The butterfly effect

See https://fourways.readthedocs.io/ section 'The butterfly effect' for a more in-depth analysis of Lorenz model. These articles provide further reading:

Étienne Ghys, 'The Lorenz attractor, a paradigm for chaos', *Chaos* (2013): 1–54, p. 20

Edward N. Lorenz, 'Deterministic nonperiodic flow', *Journal of Atmospheric Sciences* 20, no. 2 (1963): 130–41

Colin Sparrow, *The Lorenz Equations: Bifurcations, Chaos, and Strange Attractors*, Vol. 41, Springer Science and Business Media, 2012

The night sky: part 2

I tell the story in this chapter primarily based on Hamilton's own slides. Margaret H. Hamilton, 'The language as a software engineer,' Keynote (ICSE 2018) Celebrating 50th Anniversary of Software Engineering, http://www.htius.com. Further references: https://futurism.com/margaret-hamilton-the-untold-story-of-the-woman-who-took-us-to-the-moon

M. D. Holley, Apollo Experience Report – guidance and control systems: primary guidance, navigation, and control system development, National Aeronautics and Space Administration, 1976

Margaret H. Hamilton, 'What the errors tell us', *IEEE Software* 35, no. 5 (2018): 32–7

The perfect wedding

I learned about wedding planners from 'I'm a wedding planner – this is what it's like behind-the-scenes' by Tzo Ai Ang. All inaccuracies in describing a wedding planner's life are my own. https://www.newsweek.com/im-wedding-planner-this-what-like-behind-scenes-1577321

A message from B to C

For more on Betty and Claude Shannon see: https://blogs.scientificamerican.com/voices/betty-shannon-unsung-mathematical-genius/

'You had a similar example in your article, didn't you?': Claude Elwood Shannon, 'A mathematical theory of communication', *Bell System Technical Journal* 27, no. 3 (1948): 379–423

Betty Shannon did help her husband, Claude, write his later papers and worked also on the theory of entropy. The description of the dinner is, though, a fictional reconstruction of events. The example strings used in the text, and reproduced by Esther in the library, are adapted from some of the examples (on page 7) of Shannon's article.

Information equals randomness

Esther's explanation of the randomness in a cellular automaton is not complete. See https://fourways.readthedocs.io/ section 'Information equals randomness' for more details.

Twenty questions

I got some help with tricks for playing twenty questions from https://www.quora.com/What-are-the-five-most-important-questions-to-ask-in-a-game-of-20-questions

Notes and References

Entropy never decreases

Ilya Prigogine and Isabelle Stengers, *The End of Certainty*, Simon and Schuster, 1997

Word games

Claude E. Shannon, 'Prediction and entropy of printed English', *Bell System Technical Journal* 30, no. 1 (1951): 50–64

Taking the high road

'hand-washing rates after using the toilet': https://www.cdc.gov/handwashing/why-handwashing.html

'49 per cent of Brits wouldn't go, even if there is no risk': https://yougov.co.uk/topics/politics/articles-reports/2019/07/20/half-britons-wouldnt-want-go-moon-even-if-their-sa

A sea of words

Sergey Brin and Larry Page later went on to form Google. The notes from their 1998 course are still available here: http://infolab.stanford.edu/~sergey/

Class IV: Complex Thinking

The World Congress

This chapter is based on Andrej N. Kolmogorov, 'Combinatorial foundations of information theory and the calculus of probabilities', *Russian Mathematical Surveys* 38, no. 4 (1983): 29–40. The quotes are paraphrased from the original article.

The matrix

J. R. Pierce and Mary E. Shannon, 'Composing music by a stochastic process', *Bell Telephone Laboratories, Technical Memorandum MM-49-150-29* (1949)

Haizi Yu and Lav R. Varshney, 'On "Composing music by a stochastic process": from computers that are human to composers that are not human', IEEE Information Theory Society Newsletter, Vol. 67, No. 4 (2017): 18–19

The streets of London

The stories and statistics were adapted from http://www.streetsoflondon.org.uk/about-homelessness. I made a donation (and you are encouraged to make one too!) to help support the work.

I, II, III, IV

See https://fourways.readthedocs.io/ section 'I, II, III, IV' for a more in-depth analysis of cellular automata models.

See also Stephen Wolfram, *A New Kind of Science*, Vol. 5. Champaign, IL: Wolfram Media, 2002

All of the life

Mark D. Niemiec, 'Synthesis of complex life objects from gliders', *New Constructions in Cellular Automata* (2003): 55

Paul Rendell, 'Turing machine universality of the game of life', PhD dissertation, University of the West of England, 2014

Ananyo Bhattacharya, *The Man from the Future: The Visionary Life of John von Neumann*, Penguin UK, 2021

Christopher G. Langton, 'Self-reproduction in cellular automata', *Physica D: Nonlinear Phenomena* 10, nos. 1–2 (1984): 135–44

Christopher G. Langton (ed.), *Artificial Life: An Overview*, MIT, 1997

Figure 24 is based on creations by Yonaton: https://twitter.com/zozuar

Examples of complex animal behaviour: David J. T. Sumpter, *Collective Animal Behavior*, Princeton University Press, 2010

The hard edges of social reality

The examples in this chapter and, particularly, the 'segregation' at the party are inspired by Thomas C. Schelling, *Micromotives and Macrobehavior*, W. W. Norton and Company, 2006.

V-shaped pedestrian formations are studied empirically in this article. Mehdi Moussaïd at al., 'The walking behaviour of pedestrian social groups and its impact on crowd dynamics', *PloS One* 5, no. 4 (2010): e10047

'the majority of people are in all-male or all-female groups': For women, the average proportion of women in their group is

$$\frac{7\times\left(7/\left(7+7\right)\right)+5\times\left(8/\left(5+8\right)\right)+28\times1}{40}=86.4 \text{ percent}$$

For men, the average proportion of men in their group is

$$\frac{7\times\left(7/\left(7+7\right)\right)+8\times\left(8/\left(5+8\right)\right)+45\times1}{60}=89.0 \text{ percent}$$

A person is a person through other people

Desmond Tutu, 'Speech: No future without forgiveness (version 2)' (2003). Archbishop Desmond Tutu Collection Textual. https://digitalcommons.unf.edu/archbishoptutupapers/15

A good starting point for Ubuntu is Abeba Birhane, 'Descartes was wrong: "a person is a person through other persons" ', *Aeon* (2017)

Another insightful reading of Ubuntu is Nyasha Mboti, 'May the real ubuntu please stand up?' *Journal of Media Ethics* 30, no. 2 (2015): 125–47

Thank you to Anne Templeton at the University of Edinburgh for discussions about crowds. I used the following references in this section.

Dirk Helbing, Anders Johansson and Habib Zein Al-Abideen, 'Dynamics of crowd disasters: an empirical study', *Physical Review E* 75, no. 4 (2007): 046109

Hani Alnabulsi and John Drury, 'Social identification moderates the effect of crowd density on safety at the Hajj', *Proceedings of the National Academy of Sciences* 111, no. 25 (2014): 9091–6

Hani Alnabulsi et al., 'Understanding the impact of the Hajj: explaining experiences of self-change at a religious mass gathering', *European Journal of Social Psychology* 50, no. 2 (2020): 292–308

Anne Templeton, John Drury and Andrew Philippides, 'Walking together: behavioural signatures of psychological crowds', *Royal Society Open Science* 5, no. 7 (2018): 180172

David Novelli et al., 'Crowdedness mediates the effect of social identification on positive emotion in a crowd: a survey of two crowd events', *PloS One* 8, no. 11 (2013): e78983

Almost always complicated

Per Martin-Löf, 'The definition of random sequences', *Information and Control* 9, no. 6 (1966): 602–19

A life in short scenes

'. . . entire international mathematical community': Albert N. Shiryaev, 'Kolmogorov: life and creative activities', *Annals of Probability* 17, no. 3 (1989): 866–944, pp. 869–71

'. . . on a several-week-long boat trip down the Volga': Albert N. Shiryaev, 'Kolmogorov: life and creative activities', *The Annals of Probability* 17, no. 3 (1989): 866–944, p. 882

'Noether boys': Pavel S. Aleksandrov, 'Pages from an autobiography', *Russian Mathematical Surveys* 34, no. 6 (1979): pp. 297–9, 267

'. . . forever hold in his memory': Pavel S. Aleksandrov, 'Pages from an autobiography. II', *Russian Mathematical Surveys* 35, no. 3 (1980): pp. 315, 317

'small fishing village right on the coast': Pavel S. Aleksandrov, 'Pages from an autobiography. II', *Russian Mathematical Surveys* 35, no. 3 (1980): pp. 315, 318–19

'. . . over incredible distances': Paul M. B. Vitányi, 'Andrei Nikolaevich Kolmogorov', *CWI Quarterly* 1, no. 2 (1988): 3–18

'. . . effectiveness of artillery barrages': Albert N. Shiryaev, 'Kolmogorov: life and creative activities', *The Annals of Probability* 17, no. 3 (1989): 866–944, p. 907

'. . . practical view of maximum likelihood': see for example Mátyás Arató Andrei Nikolaevich Kolmogorov and Ya G. Sinai, 'Evaluation of the parameters of a complex stationary Gauss-Markov process', *Doklady Akademii Nauk SSSR* 146 (1962): 747–50

'interacting species in an ecosystem': A. Kolmogorov, 'Sulla teoria di Volterra della lotta per lesistenza', *Gi. Inst. Ital. Attuari* 7 (1936): 74–80

'. . . disturbance in the air as an aeroplane takes off': Uriel Frisch, and Andreĭ Nikolaevich Kolmogorov, *Turbulence: The Legacy of A. N. Kolmogorov*, Cambridge University Press, 1995

'. . . which entropy measured': Claude Elwood Shannon, 'A mathematical theory of communication', *Bell System Technical Journal* 27, no. 3 (1948): 379–423, p. 379

An indescribable explanation

Sokal's account of his experiment can be found here: http://linguafranca.mirror.theinfo.org/9605/sokal.html

A worthy life

This chapter is based on stories in Yu K. Belyaev and Asaf H. Hajiyev, 'Kolmogorov Stories', *Probability in the Engineering and Informational Sciences* 35, no. 3 (2021): 355–68

Some of Kolmogorov's views on mathematics education can be found here: https://mariyaboyko12.wordpress.com/2013/08/03/the-new-math-movement-in-the-u-s-vs-kolmogorovs-math-curriculum-reform-in-the-u-s-s-r/

About the Author

David Sumpter is professor of applied mathematics at the University of Uppsala, Sweden. He is the author of the international bestseller *The Ten Equations That Rule the World* as well as *Soccermatics* and *Outnumbered,* which have been translated into ten languages. He has worked with a number of the world's biggest football clubs, advising on analytics.

www.david-sumpter.com